The Radio
Noise Spectrum

Edited by DONALD H. MENZEL

Harvard University Press

Cambridge, Massachusetts

1960

Willimantic State College Library
J. EUGENE SMITH LIBRARY
EASTERN CONN. STATE UNIVERSITY
WILLIMANTIC, CT 06226

WITHDRAWN FROM

© Copyright 1960 by the President and Fellows of Harvard College

All rights reserved

Distributed in Great Britain by Oxford University Press · London

Library of Congress Catalog Card Number 60-7997

Printed in the United States of America

621.384
M52

Preface

This modern era has had many names: the golden age, the machine age, the atomic age, the electronic age, and so on. One further title, hitherto unpublicized, it eminently deserves: the age of noise. Man has compounded the natural noise that preceded his existence on the earth until no point on this globe is free from it. Even in the desert's hush, radio waves pervade the air and provide a source of potential noise.

The shorter waves escape from the earth and fill interplanetary space with the intermingled clamor of FM, TV, radar, and other insistent voices. If Martians exist, which I very much doubt, and if they possess our present scientific capabilities, they should have little difficulty in receiving and perhaps even understanding some of our radio programs. We have already bounced signals from the moon and received them again, two and one half seconds later. Scientists of the USSR have recorded TV pictures of the other side of the moon, sent from a lunar probe. Artificial satellites collect information, store it, and release it on command at appropriate moments when the satellite is near the earth.

The scientist, concerned only with his particular area of observation, strives to pick from the ether the one radiation of special interest to him. All other radiation, whether artificial or otherwise, he regards as noise. The wanted radiation may be one of Nature's emissions from the sun, planets, radio stars, or the depths of the universe, rather than radiation carrying a verbal message. Whatever the source of the emission, the amount of information that can be received in a given interval depends directly on the signal-to-noise ratio, and if scientific studies are to prosper, earnest efforts must be made to protect certain frequencies by forbidding transmissions in the useful ranges. Artificial satellites constitute an additional hazard to the preserva-

37781

tion of these channels by acting as passive reflectors or by telemetering observational data.

This book deals with the important problem of radio noise, its sources, whether man-made or natural, over the known range of frequencies. Certain of these contributions will interest the communicator, enabling him to estimate the potential interference from various types of sources. Other contributions deal mainly with scientific problems, such as the origins and significance of certain characteristic noise radiations.

The individual chapters of this book derive from papers presented by the respective authors, at a Conference on Radio Noise, held at Harvard College Observatory, April 22, 1958. I wish to express my appreciation to the Department of Defense, U.S. Army Signal Corps, for sponsoring this conference, under contract DA 49-170-sc-2386.

DONALD H. MENZEL

March 2, 1960

Contents

The Radio Noise Spectrum

1

Man-Made Radio Noise

E. W. ALLEN

Man-made noise, which may be characterized as a form of electro-magnetic refuse, shares with atmospheric and circuit noise the doubt-ful distinction of being one of the first forms of interference to communications. In treating this subject it is necessary to include conducted man-made noise, since much of the interference to radio communications arises by conduction rather than by radiation through space. And, in view of the ways in which much of the noise is generated, interest cannot be limited to the nominal boundaries of the radio spectrum but must extend to frequencies below the lower boundary of 10,000 c/s. The noise to which I refer is characterized by a broad spectrum. It is included under the general term of "interference," but is here distinguished from the single-frequency, or narrow-band, interference which is produced by radio transmitters, industrial radio-frequency generators, diathermy, and other continuous-wave generators.

The principal sources of man-made noise to be found in residential areas are power lines, rotating electrical machines (especially those with commutators), arcing electrical contacts, gaseous discharge lighting, and vehicular ignition. In industrial areas, additional contributions are found from such apparatus as electric arc welders and smoke precipitators. The noise generated by many of these devices is impulsive in nature and, although the repetition rate or the fundamental frequency of the pulse envelope may be low, the very steep wave shapes or the shock excitation of electrical circuits may produce strong components which extend far into the radio-frequency spectrum. The great variety of noises produced and the difference in relative nuisance value to various types of communication have complicated the production of reliable instruments and the accumulation of quantitative information on noise levels.

Since man-made noise is man-made, it would seem that it would be subject to ready control and would long since have ceased to be a problem. But such are human ingenuity and productivity that new devices which contribute are continually making their appearance. Many of these are designed and built by persons unaware of communications requirements, and the problem becomes apparent only after many have been sold and put into use. Or, worse still, they remain unidentified and continue to add to the din, anonymously. Also, economics plays a strong part, and a balance has to be struck between the costs of noise elimination and the advantages to be gained by the communications affected. Many devices which are known to be prolific sources of radio noise are sold today with little or no attempt at suppression because of the added cost and the lack of applicable laws.

Long before the advent of radio, the operators of early communications circuits found that the control of electrical noise was essential. Joint coordination groups were organized by the power and communications companies. These groups have established a cooperative pattern which has existed to the present day, and which has been carried over into the efforts to control the sources of radio interference, including radio noise.

In the early days of radio communication, the widespread existence of radio noise was recognized and on land was combated chiefly by choosing relatively quiet radio receiving sites in locations remote from populated areas. Special attention could then be given to the power lines entering such sites and to the electrical machinery operating nearby for the suppression of noise. It was not until after the advent of radio broadcasting in the early 1920's, when the necessity for protecting numerous receiving sites in populated areas arose, that there was a concerted effort by industry and government to attack the problem of radio noise on land.

Aboard ships, aircraft, and land vehicles, where electric power and communications equipment must operate in close proximity, adequate suppression was necessary from the beginning. Since both were under the control of the same entity, the suppression measures could be tailored to fit as closely to the needs of communication as practical. It was soon found that the practical limits of suppression depended greatly upon the awareness of the problem in the design and manufacture of components, and much of the effort, particularly in the armed services, has been devoted to the development of noise specifications for electric power components. For this purpose, standardized measurement procedures and instruments capable of reliable quantitative measurements are essential.

Although the problems have proved to be very complex, and have not readily yielded to solution, significant progress has been made over the years, principally through the joint efforts of a group of interrelated committees, the first of which dates back to 1924 and which developed chronologically as follows: National Electric Light Association Committee, National Electrical Manufacturers Association Committee, EEI-NEMA-RMA Joint Coordination Committee, IEC International Special Committee on Radio Interference (CISPR), ASA Sectional Committee C-63, and SAE-RMA Vehicular Interference Subcommittee.

These committees have worked long and faithfully in attempts to develop reliable instrumentation and test procedures. Several specifications were issued which represented milestones of progress, but it is only relatively recently that some of the problems having to do with instrumentation appear to be approaching solution.

Tests by the SAE-RMA Subcommittee indicated that the levels of ignition noise from automobiles could be reduced some 20 db or so by the use of lumped resistors of about 10,000 ohms each in the high-voltage distribution and spark-plug leads. A quantitative standard was not adopted, partly because there was about a 10-db difference in the values measured by different meters. While the tests showed a marked improvement with resistors in the high-voltage leads, their use did not become general in the automotive industry, partly because of cost and partly because of the incorrect conclusion on the part of many mechanics that their removal would improve engine performance. Authoritative tests, however, indicated longer spark-plug life without degraded performance. In 1947, tests with distributed resistance in the form of high-resistance nonmetallic conductors showed improved suppression performance at frequencies above 100 Mc/s and eliminated the need for separate resistors. While this was not universally adopted, the widespread use of radio receivers in automobiles and the concomitant need for suppression is rapidly decreasing the automobile noise problem. Trucks and busses are still found to be very noisy in many cases.

As far as I am aware, there are no recent measurements of the levels of ignition noise to be found in populous areas. Quantitative measurements on frequencies between 40 and 450 Mc/s were made back in 1940 by R. W. George, but these are likely to be no longer valid because of changes in automobile design and of the factors to which I have referred above.

In 1945 the FCC instituted proceedings to reexamine the status of aural broadcast stations operating in the medium-frequency band between 550 and 1600 kc/s. A joint industry-government committee was appointed, among whose duties was the assessment of satisfactory signal levels in the presence of man-made noise. Two studies were undertaken, the first to determine by subjective tests the signal-to-noise ratios for atmospheric and for man-made noise which were acceptable for a broadcast service, and the second to ascertain the noise levels existing in cities of various sizes.

The method of determining acceptable signal-to-noise ratios that was adopted by the committee consisted in making tests of audience reaction when listening to carefully prepared recordings of music and speech having selected signal-to-noise ratios. By holding listening tests at many locations throughout the country a sampling of about 2000 listeners was obtained which was fairly representative of the radio audience. The music tests showed ratios of 18 db acceptable to 10% of the listeners; 26 db to 50%; and 34 db to 90%. The corresponding ratios in speech tests were 21 db, 29 db, and 37 db.

Quasi-peak noise meters of the type used to measure the values of noise for the signal-to-noise records were used also to measure the noise at a frequency of 1 Mc/s at street level in communities ranging from 1000 to 3,000,000 in population. While there was a very wide scatter of the data, Table 1 shows values taken from normalized distributions. From the signal-to-noise ratios and the noise levels thus measured, the required signal levels to provide radio service in cities of various sizes can be estimated.

TABLE 1. Noise levels (db above $1\mu v/m$) exceeded for various percentages of locations.

City Population	Percentages			
	10%	30%	50%	70%
10^6	63	49	40	31
10^5	47	38	32	25
10^4	39	32	27	22
10^3	34	28	24	20

Many cities have enacted ordinances directed toward the control of radio noise and other interference. A model ordinance was prepared by the National Institute of Municipal Law Officers (NIMLO) as a guide to cities in the matter. Both the model ordinance and the early city ordinances with which I am familiar are concerned with the protection of aural broadcasting in the frequency range 540 to 1600 kc/s, or the so-called "standard broadcast" band. More recently some cities have expressed the desire to enact ordinances against television interference and the NIMLO has requested the FCC to assist it in drafting a new model. The old model is based, and presumably the new model would be based, upon legal field-strength limits, with penalties for failure to comply. Legally this seems quite simple, but technically it is not, and as a result of our study, I personally am inclined to believe that, from the technical viewpoint, a more practical method of local radio noise control would be the inclusion of reasonable suppression requirements in the electrical codes, together with an attack on the noise-producing capabilities of equipment at the manufacturing level. From what has been said above, it is apparent that the making of reliable and reproducible field-strength measurements is not easy, even under laboratory or standard field conditions, and is extremely difficult or even impractical in the normal urban situation.

While many other countries, including Canada and Great Britain, have national laws with specific limitations on the generation of radio and electrical noise, there is no such law in the United States. The nearest thing in the FCC rules is the provision of Section 15.31 which requires that incidental radiation devices shall be operated in such a manner that the radiated energy does not cause harmful interference. This is, in effect, a back-up rule which permits the FCC to take action in cases where persuasion and cooperative effort have not been effective. There are rules (Part 18) governing radio-frequency generators and electric arc-welding equipment.

The FCC plans to protect the 1400–1427-Mc/s hydrogen 21 cm line for use in radio astronomy. Other frequencies in which radio astronomers might be interested will be considered on a case-by-case basis, to give them as much protection as possible in particular areas. At the West Virginia astronomical observatory site, and other sites nearby, specific provisions are being made for considering radio assignments and other sources of interference within particular zones around each site.

The FCC has also been active in the establishment of interference-correction committees composed of industry representatives, radio amateurs, and other interested persons. Through its field organization, interest is aroused at the local level in the formation of Cooperative Interference Committees (CIC) or Television Interference Committees (TVIC), as appropriate. A Cooperative Interference Committee is a group of industrial officials who have agreed to assist the Commission in interference elimination. The CIC turns over to the FCC difficult or controversial interference cases which require official Commission decisions or enforcement procedures, but many cases can be solved by the CIC without FCC intervention. CIC organizations deal with industry or commercial radio interference problems while the older TVI committee program deals only with television interference attributed to amateur stations. At the present time there are 32 CIC organizations through the United States and territories with members from many branches of radio, television, and electronics industries. CIC organizations have solved interference problems involving radio services in aviation, maritime communications, public safety, and industry, as well as assisting military agencies in interference cases.

BIBLIOGRAPHY

Methods of measuring radio noise, NEMA Publication No. 102 (National Electrical Manufacturers Association, New York), 1935.

Methods of measuring radio noise, NEMA Publication No. 107 (National Electrical Manufacturers Association, New York), 1940.

"American standard specification for radio noise and field strength meters, 0.015 to 30 Mc/s," (American Standards Association, New York), C 63.2, November 1957.

"American standard on methods of measurement of radio influence field (Radio Noise), 0.015 to 25 Mc/s: low voltage electric equipment, and non-electric equipment," (American Standards Association, New York), C 63.4, November 1957.

V. D. Landon and J. D. Reid, "A new antenna system for noise reduction," *Proc. I.R.E. 27,* 188 (1939).

R. W. George, "Field strength of motor car ignition between 40 and 450 Mc," *Proc. I.R.E. 28,* 409 (1940).

"How to suppress radio interference," *Electrical Manufacturing,* pp. 110–128 (September 1954).

H. O. Merriman, *Radio interference from electrical apparatus and systems,* (Ottawa: Canadian Department of Transport, 1954).

F. D. Rowe, *How to locate and eliminate radio and TV interference* (John F. Rider, New York), 1954.

Proceedings of the [first] Conference on Radio Interference Reduction, December 1954 (Armour Research Foundation of Illinois Institute of Technology, Chicago).

Proceedings of the Second Conference on Radio Interference Reduction, March 1956 (Armour Research Foundation of Illinois Institute of Technology, Chicago).

Proceedings of the Third Conference on Radio Interference Reduction, February 1957 (Armour Research Foundation of Illinois Institute of Technology, Chicago).

NIMLO Model Ordinance Service (National Institute of Municipal Law Officers, 726 Jackson Place, N. W., Washington, D. C.).

2

The Aurora and Radio Wave Propagation

ALLEN M. PETERSON

1. Introduction

The auroras are luminous phenomena most often observed in polar regions and characterized by sudden and irregular appearance and by a variety of distinctive forms which usually undergo rapid changes of intensity, position, and color. The aurora is so conspicuous a phenomenon that it has attracted widespread attention and its study is as old as natural science itself.

The rare and unexpected occurrence of auroras in nonpolar regions, their wild and weird movements and flashing colors have all conspired in past ages to place the aurora high on the list of terrifying events. Some of the earliest references are found in the writings of the ancient philosophers, Aristotle (4th century, B.C.) and Seneca (1st century A.D.). Seneca describes how one night, a blood-red glow being observed in the west, the Roman cohorts were dispatched to Ostia to help extinguish the flames, everyone believing that seaport town at the mouth of the Tiber to be on fire. Seneca's descriptions of the common auroral forms are exceedingly clear and scientific, without superstitious flavor. Although many accounts have been written of auroral observations, it is difficult to capture in words the true picture of the strikingly beautiful light forms which constitute the aurora. Some of the best accounts can be found in the book *Farthest North* by Fridjof Nansen, in which he describes his polar expedition with Fram (1893–1896).

Scientists now generally believe that the aurora is produced by the entry into the upper atmosphere of positively charged atoms (mainly of hydrogen) and electrons. These streams of charged parti-

7

cles coming from the sun interact with the earth's magnetic field and are funneled toward the earth's geomagnetic poles. Thus, auroras are most frequently observed near the north and south polar regions. They are seldom seen more than 50 deg of latitude from the geomagnetic poles and are most frequently observed, not at the poles as one might think, but 23 deg of latitude (1600 miles) from them. Two rings around the earth, a few degrees of latitude (a few hundred miles) wide and 23 deg from the geomagnetic poles, are known as the "auroral zones." The geomagnetic poles are displaced about 11 deg from the geographic poles, toward Canada in the Northern Hemisphere and toward Australia in the Southern Hemisphere. The auroral zones reach lower geographic latitudes over Canada and in the waters south of Tasmania than anywhere else in the world.

Within recent years it has been discovered that the agency creating the visual display also creates ionized regions capable of reflecting and absorbing radio waves. Reflections from the aurora are potentially useful for communication purposes, but are of even more importance because of the interference they may cause in communications circuits. Since their suitability for communications is not established, and much research would have to be conducted on equipment, antennas, and system parameters before reliable use could be made of them, these reflections can be regarded, for the present at least, chiefly as a threat to communications and radar operation. Because the aurora is capable of reflecting signals over long distances, care must be taken in the assignment of operating frequencies to avoid interference between transmitters even though they are widely separated in the conventional sense. Military radars operating at frequencies below 1000 Mc/s can be expected to show auroral echoes on their screens even at locations far from the auroral zone. Targets will need to be detected among these auroral "clutter" echoes. An understanding of the nature of auroral echoes will bring about improvements in equipment design, and a knowledge of the statistical characteristics of the occurrence of such echoes will help to minimize the disruption of normal operation. When a radio wave is transmitted through the region in which an auroral display is taking place, it is weakened or absorbed and, as a result, communication services in and through the auroral zone are frequently disturbed. Examples of the disturbance to communications produced by the aurora can be found in the difficulties encountered in air-to-ground communications on the North Atlantic Air Traffic Control communications circuits which traverse the region near the heart of the auroral zone. For appreciable periods of time, up to many hours or even days, communications blackouts occur in which it is impossible to communicate to or from aircraft flying beyond the line of sight of the ground station attempting communications.

Reasons such as those discussed above account for the recent upsurge in research on the influence of the aurora on radio wave propagation.

2. *Visual and Optical Studies of the Aurora*

The visual aurora appears in many forms: quiet arcs, homogeneous bands, pulsating surfaces, and sometimes just a diffuse glow. There occur also forms with ray structure, which resemble curtains or draperies; most beautiful of all is the "corona," in which all rays or draperies converge above the observer, creating for him a celestial crown. In auroras with ray structure it is known that the rays lie parallel or nearly parallel to the lines of the earth's magnetic field. This rayed structure is shown rather well in the photographs of Figs. 1 and 2.

Much of the presently available information on the aurora has been obtained by systematic study of visual observations by volunteer observers at many locations around the world. One of the best summaries of this type of observation is that published in 1873 by H. Fritz, professor at the Zurich Polytechnic Institute. It covered the period from 503 B.C. to A.D. 1872. Fritz's work was brought up to date in 1946 by Vestine at the Carnegie Institution of Washington. The work of Fritz and Vestine has led to the preparation of charts showing the geographic distribution of auroral occurrence rates. Figure 3 is an example of such a chart for the Northern Hemisphere. The "auroral zone" is defined as the region near the line of maximum frequency of occurrence.

The auroral zones thus defined represent regions for which, on a long-term statistical basis, the visual auroral occurrence rate reaches a maximum. It should not be inferred from charts showing the auroral zones that at any given time the auroral disturbance will be a maximum in the center of the auroral zone. The shape and geographic extent of the intense regions of auroral activity are not well understood at the present time.

During the International Geophysical Year (1957–58) much additional work was done by visual observers concentrated in networks which were organized in different countries. The observations by land-based observers were supplemented by observations by crews of aircraft (who could often observe when cloud and fog obscured the sky from surface observers) and ships. The success of schemes of this kind depends largely upon the cooperation of voluntary observers.

Fig. 1. Auroral arc at College, Alaska. (Photograph by V. Hessler.)

Several optical instruments have been valuable in investigating auroras. One is an "all-sky" camera developed for auroral work by C. W. Gartlein of Cornell University. It has a 16-mm motion-picture camera pointing downward to a convex mirror, which gives an image of the entire sky and makes it possible to maintain a continuous patrol of the heavens (Figs. 4 and 5). Another invaluable aid is the auroral spectrograph, which analyzes the light from an aurora and tells us the kinds of atoms and molecules present in the atmosphere, their temperatures, the amounts of energy radiated, and something of the mode of excitation. The faintness of the aurora and the rapidity with which it changes make spectroscopy difficult, but development of the technique by Norwegian, Canadian, and United States workers has produced spectrograms of auroras extending throughout the visible spectrum and into the near ultraviolet and the near infrared.

Various attempts have been made to measure the magnitude or intensity of auroras. The Geophysical Institute at College, Alaska, has in recent years used a photoelectric photometer to measure the illumination of the sky during auroral displays. The measurement is made in a small portion of the visible spectrum which identifies the auroral light—say the green auroral line. These recordings indicate that an aurora increases the illumination of the night sky tenfold on the average, and during bursts of intensity it may brighten the sky more than a hundredfold.

3. Radio and Radar Observation of the Aurora

Radio and radar methods of investigating the aurora have within the last two decades become very valuable. A radar set does not see exactly the same thing as the eye or the camera, but it has the great

Fig. 2. Multiple auroral arc at College, Alaska. (Photograph by V. Hessler.)

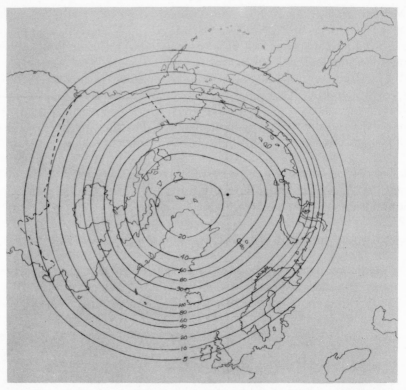

Fig. 3. Chart of auroral occurrence based on the work of Fritz and Vestine.

advantage that it can detect auroras through clouds or in daylight. Radio and radar echoes have been observed at radio frequencies between a few megacycles and a few hundred megacycles per second. These echoes are characterized by rapid fluctuations in the amplitude of the echo and relatively rapid changes in the location of the echoing region. The changes in location of the echoing region appear to result from changes in the position of the auroral forms. The rapid fading of the signal strength results from the motion of irregularities within the auroral forms. Motion of irregularities results in the shifting and spreading of the frequencies which are reflected from the aurora.

Reflections which today can be identified as echoes from the aurora were observed in the earliest days of the study of the ionosphere, but reasonable interpretation of the observations was a very slow process. Some of the effects of the aurora on radio-wave propagation were observed in 1933 by Sir Edward Appleton[1] in an expedition to Tromsö,

Norway, as a part of the program of the Second International Polar Year (the predecessor to the IGY).

Since then, phenomena believed to be associated with the aurora have frequently been observed on ionosphere sounders, which are used in a routine fashion to study the ionized regions of the upper atmosphere. Even earlier than these studies were observations in 1928 (by Hoag and Andrew[26] in connection with early short-wave radio-communications experiments) that strangely delayed signals were obtained during auroral activity.

The fact that radio echoes from the aurora were possible was known before World War II to radio amateurs[46] operating on frequencies between 30 and 100 Mc/s. Auroral radar echoes in this frequency range were first obtained by Harang[17,18] at Tromsö, Norway. After World War II, they were obtained more or less accidentally at Manchester, England, and Ottawa, Canada, by equipment erected to examine echoes from meteor trails. Since then auroral radars have been in use at a number of other locations (Kiruna, Sweden;[23] Saskatoon,

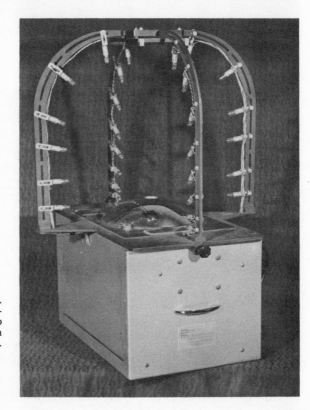

Fig. 4. All-sky camera. Single frame motion-picture camera mounted atop framework looking down on hemispherical mirror.

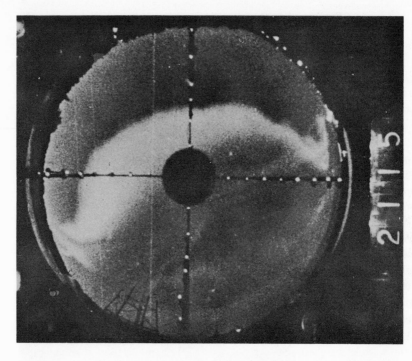

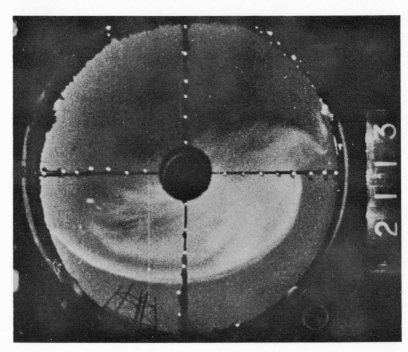

Fig. 5. All-sky camera photographs. Auroral form stretching from horizon to horizon photographed as reflection from hemispherical mirror. Note change in auroral form in 2-min interval (2113–2115 AST) between the two photographs. (Photograph by C. T. Elvey.)

14

Canada;[12,33] Ottawa, Canada;[15,34] Ithaca, New York;[11,44] Oslo and Tromsö, Norway;[20] Jodrell Bank, England;[56,31] Fairbanks, Alaska;[4,9,39] Boston, Massachusetts;[16] Stanford, California[40,41]).

4. The Geometry of Radar Reflections from the Aurora

The number of auroral radar echoes observed and their strength are markedly dependent on the orientation of the lines of the earth's magnetic field with respect to the observing location. Radar echoes are obtained primarily when it is possible to make a perpendicular reflection at a line in a region in which auroral ionization is being produced.

Waves are scattered from the aurora as if from long, thin columns of ionized gases parallel to the earth's magnetic field. As a result, radar echoes are not seen with equal ease in all directions. In fact, at Northern Hemisphere stations echoes are seen only when the antenna beam is directed toward the north. A sketch showing the geometry involved is presented in Fig. 6. Strong echoes are seen only when the ray from the transmitter to the auroral ionization meets the earth's magnetic field line at or near perpendicular incidence. The requirement for perpendicular incidence is not an absolute one and is a function of frequency. It has been found that a radar operating on a frequency near 400 Mc/s sees echoes when the ray from the transmitter meets the magnetic field line at angles up to 7° or 8° from the perpendicular. At lower frequencies, larger departures (about 15° at 100 Mc/s) are observed, while at higher frequencies the departures are smaller (about 4° at 800 Mc/s).

Figures 7, 8, 9, and 10, which depict the influence of magnetic-field geometry on the regions from which auroral echoes can be expected to occur, are based upon observations at Stanford, California; Seattle, Washington; and College, Alaska. In Fig. 7 for Stanford (43° geomagnetic latitude), it can be seen that in the magnetic north direction per-

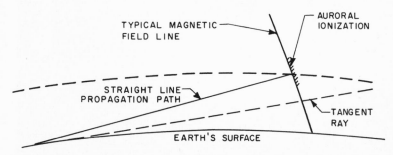

Fig. 6. Sketch illustrating earth's magnetic field lines and geometry of auroral reflection.

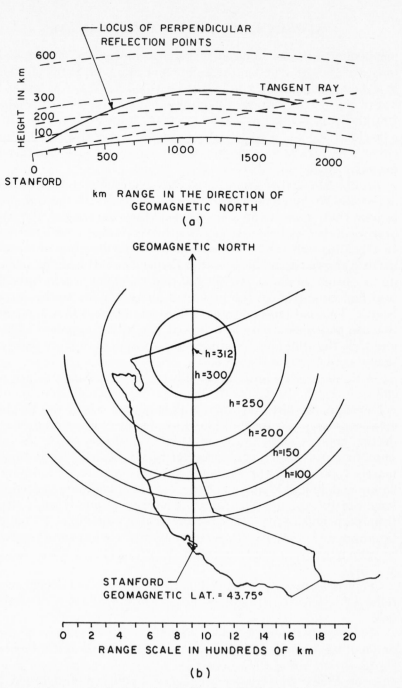

Fig. 7. Loci of perpendicular reflection for Stanford, California; (a) Locus of points of perpendicular reflection in geomagnetic north. Geometry is illustrated in Fig. 6. (b) Loci of points of perpendicular reflection contours for a succession of heights (h) above the earth's surface.

pendicularity can be achieved at heights in excess of 300 km and at ranges of 1600 or 1700 km to the north of Stanford. Perpendicularity at E-layer height (100 km) is first obtained at ranges of 200 or 300 km to the north. Perpendicular reflection can occur at a height of 100 km in directions as much as $\pm 60°$ from magnetic north. At a height of 300 km, perpendicularity can be achieved within the small circle labeled H = 300, and hence occurs only for directions close to magnetic north.

In Fig. 8, a similar geometric sketch is presented for observations at Seattle, Washington (54° geomagnetic latitude). In this case, it can be seen that, because the magnetic field lines are steeper, the region over which perpendicularity can be achieved is much more restricted. In Fig. 8 (*a*) a sketch for transmissions in the direction of magnetic north is shown. It can be seen that perpendicularity can be achieved up to a height of about 180 km and that the point at which the contour first cuts through the E-layer is about 400 km to the north of Seattle. The contours of Fig. 8 (*b*) illustrate the fact that in no direction can perpendicularity be achieved at a height as great as 200 km and that the 100- and 150-km height contours are more restricted than was the case for Stanford.

A similar sketch for College, Alaska (65° geomagnetic latitude), shown in Fig. 9, illustrates the fact that perpendicularity cannot be achieved in any direction, including magnetic north, at a height as great as 100 km. Figure 9 (*a*) illustrates the locus of perpendicular reflection points looking in the direction of magnetic north; the maximum height at which perpendicularity can be achieved is approximately 75 km. Figure 9 (*b*) shows the perpendicularity contours for 25 km and 50 km. Figure 10 shows a series of contours plotted for a height of 100 km and illustrates the location of points which depart from perpendicularity by increasing angles, starting at 1° and continuing up to 7°. These "off-perpendicularity" contours will be referred to again later in our discussion of the location of radar echoes received at College.

Figures 7 through 10 have illustrated the marked dependence of radar echo characteristics upon the geometry of the earth's magnetic field.

The earth's magnetic field is similar to that of a magnetic dipole or small bar magnet located near the center of the earth. Figure 11 shows a number of field lines in a sector extending from pole to equator. Line *a* through the geomagnetic pole is a straight line perpendicular to the earth's surface. At the equator, as illustrated by line *f,* the field lines are parallel to the earth's surface. Field lines close to the pole, such as *b,* are nearly vertical (large dip angle). Lines *c, d, e,*

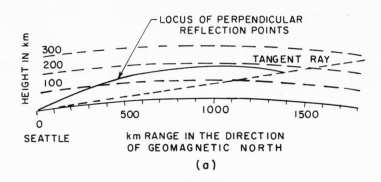

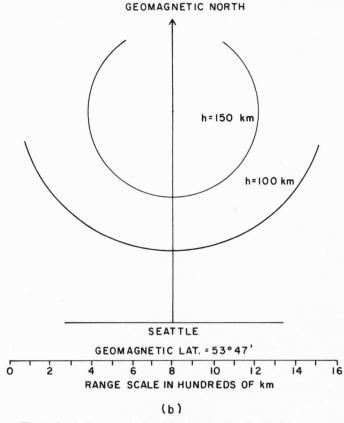

Fig. 8. Loci of perpendicular reflection for Seattle, Washington.

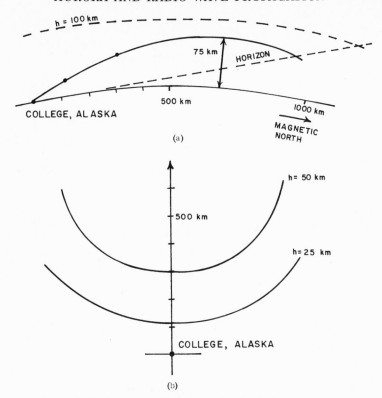

(a)

(b)

Fig. 9. Loci of perpendicular reflection for College, Alaska.

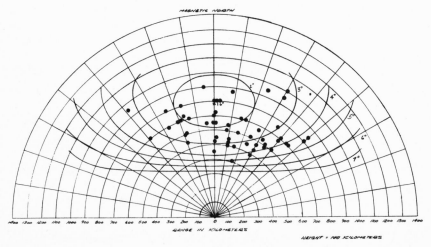

Fig. 10. Reflection geometry for College, Alaska, 24 March 1957, showing contours at a succession of angular degrees of departure from perpendicular reflection. Reflection is assumed to occur at 100 km height above the earth.

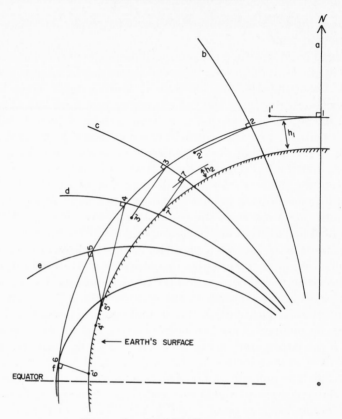

Fig. 11. Details of field line geometry.

and *f,* which are successively further from the pole, are more and more inclined from the vertical (decreasing dip angles).

Also shown in Fig. 11 are a group of straight lines (rays) which are perpendicular to the field lines at height h_1 above the earth's surface. Rays such as 1-1′, 2-2′, and 3-3′ are perpendicular to field lines of small dip angles and hence do not intersect the earth. Rays such as 4-4′, 5-5′, and 6-6′ are perpendicular to field lines of larger dip and hence do intersect the earth's surface at points 4′, 5′, and 6′. For a given height above the earth's surface, such as h_1 of Fig. 11, there will always be some point, such as point 4, which is the point closest to the pole from which a perpendicular to the field line will intersect the earth. This point (point 4 for height h_1) is the closest to the pole at which it is possible to obtain optimum radar reflection conditions for height h_1. The ray from this most poleward point is tangent to the earth at 4′ in Fig. 11. Thus for Fig. 11, a radar located at point 4′ would direct its beam northward at zero elevation (tangent to the earth at point 4') to obtain a reflection at perpendicular incidence to field line d at point 4.

Though point 4 is the most poleward perpendicular reflection point for height h_1, it may be seen that the corresponding radar location, point 4', is not the closest to the pole at which a radar can be located in order that a ray from the radar may meet a field line at a right angle. Line 5–5' illustrates that a ray from a radar at point 5', which is north of point 4', can intersect field line e at height h_1 at right angles at point 5. It should be noted, however, that reflection point 5 is considerably south of reflection point 4. It should also be noted that ray 5–5' meets the earth at a larger angle than does ray 4–4'.

If a lower reflection height is chosen, such as h_2 in Fig. 11, then it is found that a ray from the earth's surface (such as ray 7–7') can meet a field line at right angles in a region closer to the pole.

It is possible to plot on a map the locus of points of most northerly reflection at perpendicular incidence to a given field line for a given height h. This locus is composed of points such as point 4 of Fig. 11 projected downwards to the earth's surface. The corresponding locus of radar positions on the ground (such as point 4' of Fig. 11) can also be mapped. This has been done in Fig. 12 for a reflection height h = 100 km, and in Fig. 13 for h = 200 km. Included in Figs. 12 and 13 is an area locus line showing the most northerly radar locations (such as point 5' of Fig. 11) at which a ray can make a perpendicular reflection to a field line and the corresponding locus of reflection points (such as point 5 of Fig. 11). The loci of Figs. 12 and 13 are constructed geometrically from a knowledge of the measured dip angle of the actual earth's field, rather than for an assumed dipole field.

The interpretation of radar echoes is more difficult than the interpretation of the optical aurora. Not only is the occurrence of the aurora required in the region which the radar is surveying, but also geometric considerations must be taken into account. Thus, the radar auroral zone can be expected to depart markedly from the optical zone (defined as the region of most frequent occurrences of overhead visual auroras). The optimum zone for radar echoes is a combination of the optical zone of maximum occurrence and the region of favorable geometric conditions for radar reflection. The most favorable condition for observing radar echoes from auroral ionization at a given height above the earth will occur when the perpendicular reflection loci (shown by solid lines in Figs. 12 and 13) correspond closely to the region of maximum visual auroral activity (shown as a wide shaded contour line in Figs. 12 and 13). Thus, it can be seen that radar locations in the south of Alaska are very favorable for observing radar

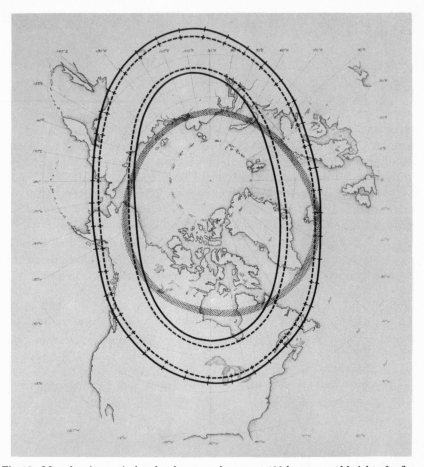

Fig. 12. Map showing optical and radar auroral zones at 100 km assumed height of reflection. *Unbroken line:* locus of most northerly aurora to which a perpendicular can be drawn that meets the earth; *unbroken crossed line:* locus of radars which, at zero elevation angle, see perpendicular to the most northerly aurora; *broken crossed line:* locus of most northerly radars capable of observing aurora perpendicular to the field lines; *broken line:* locus of positions of aurora seen by the most northerly radar capable of observing at perpendicular incidence; *shaded area:* the auroral zone.

echoes from ionization at a height of 100 km. Similarly, favorable locations exist in Norway and northwestern Russia. Somewhat less favorable conditions occur in most of southern Canada and northern United States; however, northeastern United States and southeastern Canada are regions from which nearly optimum results can be achieved over a portion of the auroral zone. The only location from which nearly perpendicular reflection can be achieved near the center of the auroral zone for a height of 200 km is the region near northern England and

Scotland and a portion of the Atlantic Ocean between Newfoundland and the British Isles.

5. *Stanford Research Institute Auroral Radar Observations*

Until relatively recent years, auroral observations have been carried out at frequencies in the HF and lower VHF band, using equipments with relatively nondirective antennas, capable of revealing only broad general details of radar reflections from auroral forms. Stanford Research Institute has recently designed high-sensitivity equipment[28] especially for the study of auroral radar characteristics. At the

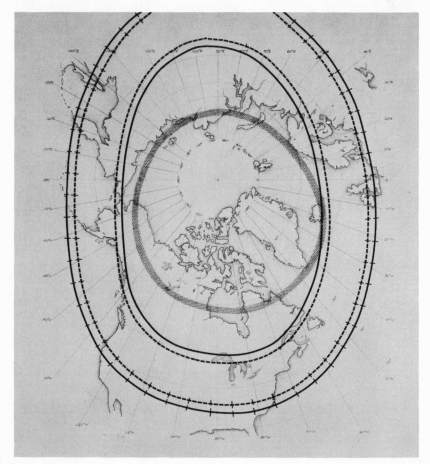

Fig. 13. Map showing optical and radar auroral zones at 200 km assumed height of reflection. (For explanation of lines, see Fig. 12.)

time of writing this equipment was operating at frequencies in the 100–800-Mc/s range. The antennas use large parabolic reflectors (61-ft diameter). Figures 14 and 15 show two such installations. The first is located in the hills behind Stanford University near Palo Alto, California. The second is located at College, Alaska, near Fairbanks, and is operated cooperatively with the Geophysical Institute of the University of Alaska. The characteristics of the radars developed for use with these antennas are shown in Table 1 for frequencies of 100, 200,

Fig. 14. Parabolic antenna, 61 ft in height, located at Stanford, California.

Fig. 15. Parabolic antenna, 61 ft in height, located at College, Alaska.

400, and 800 Mc/s. The equipment at Palo Alto is operated at 100 Mc/s, while that at College has been operated at 200, 400, and 800 Mc/s. Of special importance is the sharp beam produced by the large antennas. Beamwidths of 12° are achieved at 100 Mc/s, 6° at 200 Mc/s, 3° at 400 Mc/s, and 1½° at 800 Mc/s. The very narrow beam widths achievable at 400 and 800 Mc/s are particularly valuable in localizing the position of occurrence of reflections from the aurora. With the use of the large antennas and the higher frequencies, it has been possible for the first time to determine reflection height and position from angle and range measurements made using the radars. Examples of radar reflections obtained with the equipment at Fairbanks operating on a frequency of 400 Mc/s are shown in Figs. 16, 17, 18, 19, and 20. Figure 16 is an "A-scan" photograph of an auroral echo. Figures 17 and 18 are azimuth—range scans for fixed elevation angles. These particular photographs are for the case of discrete auroral echoes such as that obtained from an auroral arc lying to the north of the radar. Figure 19 is an elevation scan for a discrete echo of the type detected by azimuth scan in Figs. 17 and 18. The photograph in Fig. 20 shows an example of a different sort of auroral echo which we have called a "diffuse" echo. This is the same kind of display as that of Fig. 19, that is, elevation angle versus range, with intensity modulation. But this is an example of observations made in the morning when the E-layer is beginning to be illuminated by the sun. Typically, when this happens following a disturbed night (one with intense auroral activity), it is found that auroral echoes are obtained over a wide range of elevation angles and azimuths corresponding apparently to a situation in which the whole region of the sky to the north of Fairbanks is filled with field-aligned scatterers. These scatterers are capable of scattering an echo back to the radar whenever there is perpendicular reflection at the magnetic-field line. Diffuse echoes have been observed only during daylight hours. Figure 21 shows the wide extent of the region of field-aligned scatterers, typical of periods during which diffuse echoes are observed. Figures 22 and 23 show auroral observations for a frequency of 780 Mc/s. In this case, the auroral forms which produce echoes are even more localized than for 400 Mc/s, and it is sometimes found that for discrete echoes a movement of the antenna by as much as 1 or 2 deg will result in the complete disappearance of the echo.

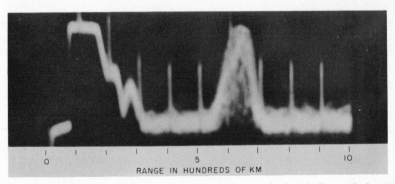

RANGE IN HUNDREDS OF KM

Fig. 16. Amplitude-range display, or A-scope, of an auroral echo. College, Alaska, 2255 AST 24 March 1957; frequency, 400 Mc/s; antenna bearing, 34° true; antenna elevation, 10°; PRF, 150 c/s.

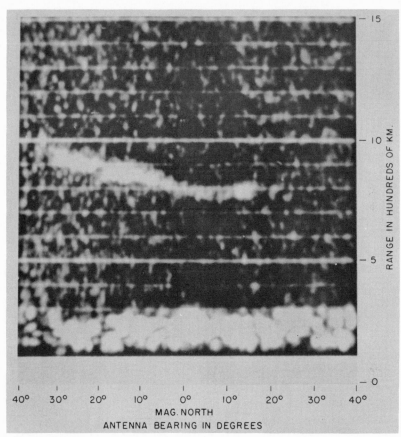

Fig. 17. Range-azimuth display, or B-scan, of an auroral echo. College, Alaska, 2220 AST 7 April 1957; frequency, 400 Mc/s; antenna elevation, 4°; sweep time, 30 sec.

Diffuse echoes are also observed during the daylight hours at 780 Mc/s.

During large ionospheric and magnetic disturbances, which occur during periods when the sun is disturbed, radar echoes are also observed at Stanford University, far to the south of the auroral zone. The observations to date have been at a frequency near 100 Mc/s, or lower; Figs. 24 and 25 show examples of echoes observed with the 106-Mc/s radar at Stanford, California. Perhaps the most important observations at 100 Mc/s have been those for which the reflection heights are found to be as great as 200 or 300 km above the earth's surface. Figure 7 demonstrates the fact that perpendicular reflection could be made at heights as great as 300 km at Stanford. The observation of radar echoes at such heights appears to demonstrate the validity of the perpendicular-reflection requirement and to illustrate the effect of reflection geometry on the appearance of radar echoes.

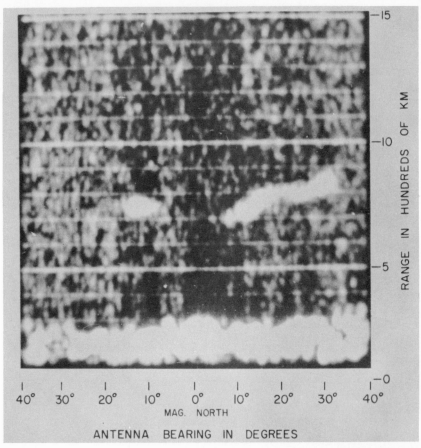

Fig. 18. Range-azimuth display, or B-scan, of an auroral echo. College, Alaska, 2245 AST 7 April 1957; frequency, 400 Mc/s; antenna elevation, 5°; sweep time, 30 sec.

Though much more intense and more frequently observed auroras are seen at College, Alaska, the heights of observations have all been in the vicinity of 100 km, corresponding to the region for which perpendicular reflection can most nearly occur. However, on those relatively rare occasions when strong auroras are seen as far south as Stanford, it is found that heights as great as 300 km are observed. The reflections occur from those regions at which perpendicular reflection is possible at great heights.

Among other characteristics of radio reflections from the aurora are rapid fading and large Doppler shifts associated with the scattering of signals from auroral forms. Figures 26 and 27 are examples of spectrum analysis of a continuous-wave signal at 400 Mc/s which is scattered from auroral ionization. It can be seen that the signal reflected from the aurora in the case of Fig. 26 is shifted up to 3 or 4

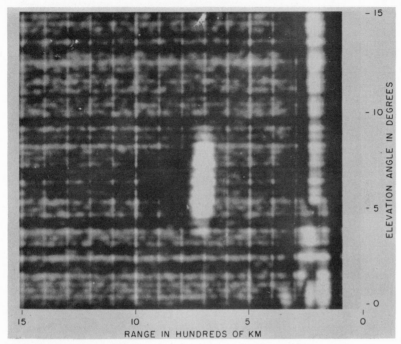

Fig. 19. Elevation-range display of an auroral echo. College, Alaska, 2342 AST 7 April 1957; frequency, 400 Mc/s; antenna bearing, 57° true; sweep time, 20 sec.

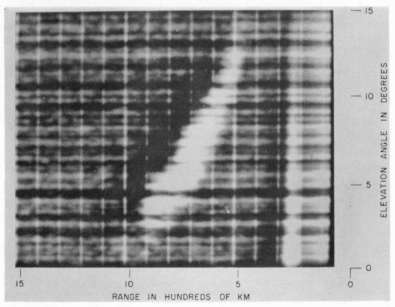

Fig. 20. Elevation-range display of a diffuse auroral echo. College, Alaska, 0711 AST 8 April 1957; frequency, 400 Mc/s; antenna bearing, 45° true; sweep time, 20 sec.

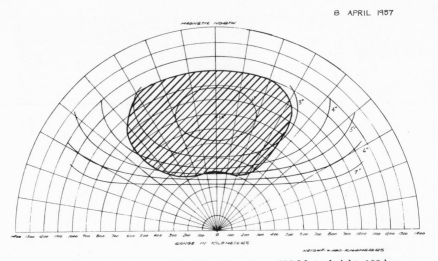

Fig. 21. Extent of a diffuse echo region. Frequency, 398 Mc/s; height, 100 km.

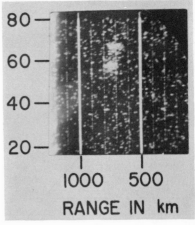

Fig. 22. Azimuth-range display of a diffuse auroral echo. College, Alaska, 0700 AST 25 March 1958; frequency, 780 Mc/s; τ, 900 μsec; elevation, 6°; PRF, 75. Vertical scale, azimuth (deg).

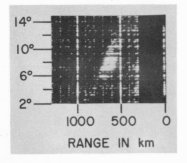

Fig. 23. Elevation-range display of a diffuse auroral echo. College, Alaska, 0630 AST 25 March 1958; τ, 900 μsec; azimuth, 33°; PRF, 75. Vertical scale, elevation (deg).

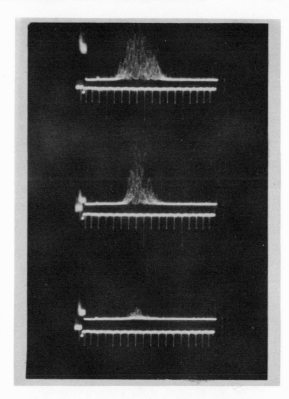

Fig. 24. Amplitude-range display of auroral echoes. Stanford, California, 0027 PST 11 February 1958; frequency, 106 Mc/s; azimuths (top) 326°, (center) 316°, (bottom) 311°; elevation, 2°.

kc/s from the frequency of the signal that is transmitted toward the auroral form. The Doppler shift in this case is toward lower frequencies, indicating that the motion of the scatterers is away from the observing radar. The spectra of Fig. 27, on the other hand, for 0910 in the morning, show shifts toward an increased frequency, indicating motion toward the observing radar. At 0911 and 0912 the scattered spectra are approximately centered on zero, but spread plus or minus several kilocycles per second. It is believed that irregularities of electron density drift in an east or west direction along the auroral zone and that the projection of this drift velocity along the ray from the radar to the reflecting point corresponds to the measured Doppler shift. Observations of this kind have been made at frequencies of 40 and 100 Mc/s by Nichols at the University of Alaska. Nichols found a close correspondence between the current systems in the E-layer of the ionosphere required to produce magnetic fluctuations on the ground, and the drifts required to produce the Doppler shifts observed with radar. Observations at 400 Mc/s have shown that the direction of drift can reverse within a matter of a few minutes from eastward to westward.

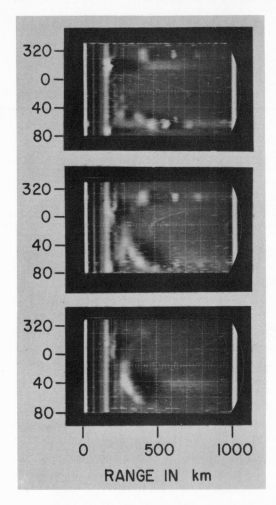

Fig. 25. Azimuth-range display of auroral echoes. Stanford, California, 2030 PST 15 March 1958; frequency, 106 Mc/s; τ, 1 ms; PRF, 150; elevations (top) 10°, (center) 18°, (bottom) 22°. Vertical scale, azimuth (deg).

6. Frequency Dependence of Echo Intensity, Aspect Sensitivity, and the Mechanism of Scattering

A study of characteristics of radar echoes in the frequency range from a few megacycles per second up to the present upper limit near 800 Mc/s should permit the formation of a model which will adequately explain the mechanism of scattering from auroral ionization. The characteristics of most importance in the formulation of a model appear to be the variation of intensity of scattered signal with frequency and the variation of aspect sensitivity with frequency.

Serious consideration has been given from time to time to models requiring "critical-density" reflection from auroral forms. Such models

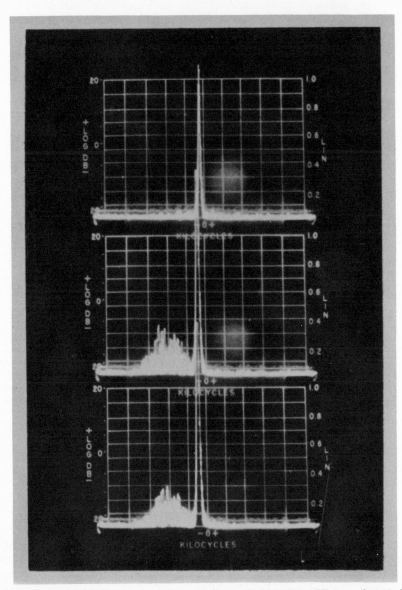

Fig. 26. Doppler display of a radar echo. College, Alaska, 0736 AST 7 April 1958; frequency, 400 Mc/s; (top) noise, (center) aurora, (bottom) aurora; azimuth, 60°; elevation, 8°. Transmitted frequency corresponds to the large spike in the center of the photograph. The Doppler-shifted echo is spread to the left of the transmitted frequency. Frequency shift indicates lower frequency and motion of scatterers away from the observing station.

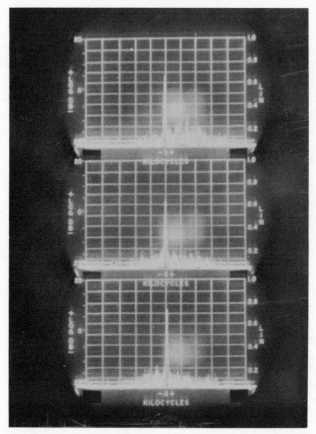

Fig. 27. Doppler display of radar echoes. College, Alaska, (top) 0910 AST, (center) 0911 AST, (bottom) 0912 AST 14 April 1958; frequency, 400 Mc/s; azimuth, 20°; elevation, 6°.

would involve plasma frequencies in excess of the observed reflection frequencies ($\geq$800 Mc/s) and, hence, very great electron densities ($\approx10^{10}$/cm³). These models do not appear reasonable on physical grounds, and, furthermore, do not appear necessary to explain the observed radar echo characteristics.

An alternative model developed by Booker, which more nearly explains the observations, involves the concept of scattering at nonisotropic irregularities[2] in electron density. These irregularities provide discontinuities in dielectric constant which are sharp compared to the wavelength of the radio wave, and hence scatter a small portion of the wave backward toward the radar, while most of the energy proceeds onward in its original direction. To explain the aspect sensitivity of the echoes, the irregularities must be in the form of columns of ionization with their long axes parallel to the earth's magnetic field.

The length of the irregularities in the field direction determines the aspect sensitivity of the scattered signal strength. An isotropic irregularity scatters equally well in the backward direction when a radio wave strikes at any angle, while a very long, thin irregularity scatters effectively in the backward direction only when the wave strikes it at right angles to the long axis. Columns of shape intermediate between these two extremes will scatter in an aspect-sensitive manner, and the allowable departure from perpendicular incidence will increase as the length decreases. Booker[2] obtained the following relation for back-scatter from elongated field-aligned irregularities:

$$\sigma_B = \frac{2^{3/2}\pi^3}{\lambda^4{}_N} \left\langle \left(\frac{\Delta N}{N}\right)^2 \right\rangle T^2\, L\, \exp\left[-\frac{8\pi^2 T^2}{\lambda^2}\right] \exp\left[-\frac{8\pi^2 L^2}{\lambda^2}\psi^2\right]$$

where L is the correlation distance along the axis of symmetry, T is the correlation distance transverse to the axis of symmetry, λ is the wavelength of the radio wave, λ_N is the plasma wavelength, ψ is the angle between the direction of wave travel and the normal to the longitudinal axis of irregularity, and $\langle(\Delta N/N)^2\rangle$ is the mean-square fractional deviation of electron density. In the derivation of this formula a Gaussian autocorrelation function was assumed for $(\Delta N/N)$. The formula gives the back-scattering per unit solid angle, per unit incident power density, per unit volume, provided the direction of incidence is nearly perpendicular to the axis of symmetry.

From observations in the 20–100-Mc/s frequency range, Booker concluded that the length $2\pi L$ of irregularities was about 40 m. The Stanford Research Institute observations at frequencies near 100, 200, 400, and 800 Mc/s indicate that echoes are observed only when the angle of incidence of the radar wave on the irregularities is within the following limits: 100 Mc/s: 10–15°; 200 Mc/s: 10–12°; 400 Mc/s: 7–9°; 800 Mc/s: 3–6°.

The aspect sensitivity found in the Stanford Research Institute experiments in the 100–800-Mc/s frequency range appears to indicate that the characteristic length of the irregularities should be reduced to about $2\pi L = 20$ m for this frequency range. If this is done, the theory outlined by Booker explains rather well the observed aspect sensitivity.

According to Booker's theory, the irregularities of electron density responsible for the scattering are created by atmospheric turbulence and made nonisotropic by the action of the earth's magnetic field. Other factors which may well play a part in creating the observed irregularities are the charged particles which produce the auroral ionization by following along magnetic-field lines and the electric fields which cause current systems to flow in the upper atmosphere during auroral disturbances.

Booker concluded that during auroras the mean electron density in the E-region was about $10^6/\text{cm}^3$ and that a mean-square deviation of electron density $\langle(\Delta N/N)^2\rangle$ of about 3×10^{-7} would explain the strength of the echoes in the 20–100-Mc/s region. He further concludes that the correlation distance transverse to the column is about $T = 0.16$ m or $2\pi T = 1$ m. Nichols, on the basis of recent experiments at 41 Mc/s and 106 Mc/s in Alaska concludes that $\langle(\Delta N/N)^2\rangle$ must be about 6×10^{-4} in order to explain the observed signal strengths. This value, although considerably higher than that originally estimated by Booker, does not seem unreasonable.

Booker's value of 0.16 m for the transverse correlation distance T was based upon an assumed upper cutoff frequency of radar echoes of about 300 Mc/s. Recent observations indicate that the cutoff frequency is above 800 Mc/s, and hence a new value for correlation distance must be assumed to explain these observations. Reasonable agreement of theory and experiment appears possible in this frequency range if it is assumed that $2\pi T \cong 0.6$ m or $T \cong 0.1$ m. Thus the ionization irregularities responsible for the observed radar echoes in the 100–800-Mc/s frequency range might consist of columns which on the average are of longitudinal correlation length $L = 3.5$ m and transverse correlation distance $T = 0.1$ m, and whose mean-square deviation in electron density $\langle(\Delta N/N)^2\rangle$ is about 6×10^{-4} in an auroral E-region of mean electron density about $10^6/\text{cm}^3$.

At College, Alaska, using the experimental radar whose parameters are shown in Table 1, measurements of the scattered signal intensity have been made at 216, 398, and 780 Mc/s. The observations have been carried out on a single frequency at a time, and simultaneous comparison of auroral echo strengths at two or more frequencies has not been possible. However, estimates of the maximum echo strengths normally encountered at the three frequencies have been made. Table 2 summarizes the results. For comparison purposes these signal-to-noise ratios were normalized with respect to the radar parameters given in Table 1. The normalized received powers, using 398 Mc/s as the reference, are shown in Table 3.

It has been assumed in this comparison that in the case of discrete echoes the scattering region is smaller than the area illuminated by the antenna, and that for diffuse echoes the scattering region is larger than the area illuminated by the antenna. Under these assumptions it may be seen that the scattered signal appears to drop off much more rapidly in the frequency from 400 to 800 Mc/s than it does in the 200- to 400-Mc/s range.

TABLE 1. Radar parameters.

Parameters	Frequency (Mc/s)			
	106	216	400	780
Peak power (kw)	50	35	40	20
Antenna gain (db) (61-ft diameter steerable parabolic reflector)	24	30	36	42
Antenna beamwidth (deg) (between half-power points in horizontal and vertical planes)	12	6	3	1.5
Receiver noise figure (db)	2.5	8	6.5	8
Receiver bandwidth (kc/s)	3–15	6	0.3–15	6
Minimum detectable signal (10^{-16} w)	1	2	1	2
Minimum detectable equivalent target cross section at 500 km (m^2)	0.4	0.3	0.03	0.03
Pulse width variable (μs)	50–2.5×10^6 and CW	450–900	400–900	450–900
Pulse repetition frequency (sec^{-1})	1500-⅓ and CW	75–150	75–150	75–150

The estimates made above of the variation of auroral echo power with frequency are rough and need refining by simultaneous multiple frequency measurements. Such measurements should be conducted with antennas of constant beamwidths for all frequencies as well as with antennas of varying beamwidths at a given frequency. In this manner, it should be possible to determine the extent of the scattering volume which contributes to the auroral echoes.

A detailed comparison of theory and experiment does not appear justified until more reliable estimates of frequency dependence of scattered power are available. However, it may be noted that in Booker's theory, which was discussed above, the wavelength dependence of the scattering coefficient σ_B is contained in the two exponential terms $\exp[-8\pi^2 T/\lambda]$ and $\exp[(-8\pi^2 L^2/\lambda^2)\psi^2]$ and that the first of these predicts a very rapid cutoff in scattered power for wavelengths shorter

TABLE 2. Scattered signal intensity.

Frequency (Mc/s)	Signal-to-noise ratio (db)	
	Discrete echo	Diffuse echo
216	30	26
398	35	22
780	13	8

TABLE 3. Normalized received power ratios.

Frequency (Mc/s)	Diffuse echoes	Discrete echoes
216	4.0	2.1
398	1.0	1.0
780	0.1	0.005

than about $\lambda = 2\pi T$. This term is the only one which is effective if the wave strikes the scattering column at perpendicular incidence. However, if wave direction departs from perpendicular, then the term $\exp[(-8\pi^2 L^2/\lambda^2)\psi^2]$ begins to cause a rapid cutoff with decreasing wavelength. Since for a radar at College, Alaska, perpendicular incidence is achievable only at heights well below the region at which most auroral ionization is produced, care must be taken to include the aspect sensitivity in any analysis of the time-frequency dependence of the scattering coefficient of auroral ionization. For irregularities of size $T = 0.1$ m and $L = 3.5$ m, a rapid cutoff in back-scattering would be expected above 400 Mc/s, and the measured aspect sensitivity could be accounted for if Booker's theoretical treatment is assumed.

7. Radio Noise Generated in the Aurora

Radio-frequency noise[22] from the aurora was reported by Forsyth, Petrie, and Currie in 1949 at a frequency of 3000 Mc/s. Covington, at about the same time (1947, 1950), also reported radiation near 2800 Mc/s from the sky, which was associated with geomagnetic and ionospheric disturbances. However, further investigations by others (Chapman and Currie in 1953; Harang and Landmark in 1953) failed to show evidence for noise emissions from the aurora. Chapman and Currie suggested that auroral noise might occur only at times of high sunspot activity.

During the recent period of high sunspot activity, a number of additional observations appear to lend considerable support to the validity of noise generation within the aurora. Hartz, Reid, and Vogan reported auroral noise observations on several occasions during 1956 at frequencies of 32, 50, and 53 Mc/s. Hartz reported a particularly striking example of radio noise emissions lasting about 1 hr at a frequency of 500 Mc/s during a large type-A red aurora, which occurred on October 21–22, 1957. Noise apparently associated with auroral activity has also been identified on the Stanford IGY back-scatter sounding records at 30 Mc/s from a number of widely separated geographic locations.

Although much more experimental evidence will be required before the characteristics of auroral noise can be understood, it now appears that positive evidence for its occurrence has been obtained. Frequency dependence, directional dependence, and polarization characteristics should all be measured in order to develop clues to the mechanism of generation of noise. Forsyth, Petrie, and Currie have suggested that plasma oscillations of the ionized volume was a possible source of noise, and Hartz has suggested that Čerenkov radiation is a more likely cause. Another possible mechanism which should be investigated is synchrotron radiation from high-energy electrons spiraling in the earth's magnetic field.

Discussion

C. O. Hines, *Defense Research Board of Canada:* I should like to inject a note of caution into the application of the aspect-sensitivity observations. Collins and Forsyth have just completed a study of auroral scattering on bistatic paths in Canada, and have revealed the existence of at least three distinct types of auroral scatter. One of these is highly aspect-sensitive, of the type described by Dr. Peterson; one is only weakly aspect-sensitive, and one exhibits no aspect-sensitivity at all. The latter becomes more important as one goes to higher latitudes, to higher sensitivities, or to forward-scatter as opposed to back-scatter. The observations were made in the neighborhood of 40 Mc/s, and the aspect-sensitive scattering appears to be more frequency-sensitive than the other.

A. M. Peterson: There are a variety of sporadic-E-type reflection mechanisms. Among them may be the ones which you say are not aspect-sensitive at all but are associated with the aurora. We find very close correlation, for example, between a nighttime variety of sporadic-E and some of the field-aligned scatter observed toward the north where one obtains a perpendicular reflection. Also, however, cases of scattering in the 40-Mc/s region apparently associated with the aurora have been recorded from virtually all directions within the auroral zone. McNamara of the National Research Council in Canada has suggested that this happens quite frequently.

Hines: The non-aspect-sensitive scatter seems to be less frequency-sensitive than the other.

Peterson: Yes. The diffuse echoes that I referred to are much less frequency-sensitive—at least in the UHF range—than the other variety, although they too are markedly aspect-sensitive—that is, one sees echoes only from points in the region at which one has perpendicular reflections. These are high-latitude auroral phenomena.

O. E. H. Rydbeck, *Chalmers University, Sweden:* You mentioned that the angle effect depends on frequency. Is there any change in the altitude distribution which depends particularly on frequency?

Peterson: I suspect there is probably a frequency sensitivity associated with this. But we have not as yet been able to make simultaneous measurements at widely spaced frequencies at low enough latitudes to meet the conditions necessary for the strongest echoes, namely perpendicular reflection at the higher height. This will have to be done at lower latitudes. We get the reflections only at times of fairly large disturbance; at Stanford we had possibly a half-dozen cases during the high sunspot winter of 1957–58 which gave us echoes at that low latitude.

I. G. Raudsep, *The Martin Company:* With respect to the Doppler shift produced by the aurora, you mentioned that the Doppler spectrum extended over 2 to 4 kc/s at 400 Mc/s. At any given time, is the Doppler shift unidirectional or is it observed on both sides of the carrier, and is this characteristic predictable?

Peterson: Sometimes it is on one side and sometimes on the other, and sometimes it is on both sides. Observations have been made at frequencies of about 40 to 500 Mc/s. The 500-Mc/s observations were conducted by Forsyth and his group in Canada. Typically, if you average over a fair period of time, you find a tendency for a drift from east to west before midnight and west to east after midnight. However, in our 400-Mc/s sharp-beam experiments at Fairbanks we found cases in which the east-west direction of drift reversed in a matter of minutes, going from a shift of 2 or 3 kc/s on one side of the carrier to an equal shift on the other side in about 20 min.

G. E. Milburn, *U.S. Army Signal Communication Engineering Agency:* Have your studies indicated any completely predictable correlation between the intensity of solar disturbances and the intensities of the reflected signals?

Peterson: No, not a completely predictable correlation. The large disturbances seem now to be pretty well traceable to solar flares of a special kind; but the night-to-night sort of thing observed near the auroral zone still has many unpredictable characteristics.

BIBLIOGRAPHY

1. Appleton, E. V., and R. Naismith, "Scattering of radio waves in the polar regions, *Nature 143,* 243 (1939).
2. Booker, H. G., "A theory of scattering by nonisotropic irregularities with application to radar reflections from the aurora," *J. Atm. Terrestrial Phys. 8,* 204 (1956).
3. Booker, H. G., "The use of radio stars to study irregular refraction of radio waves in the ionosphere," *Proc. I. R. E. 46,* 298 (1958).

4. Bowles, K., "The fading rate of ionospheric reflection from the aurora borealis at 50 Mc," *J. Geophys. Research 57,* 191 (1952).

5. Bullough, K., and T. R. Kaiser, "Radio reflections from aurorae," *J. Atm. Terrestrial Phys. 5,* 189 (1954).

6. Bullough, K., T. W. Davidson, T. R. Kaiser, and C. D. Watkins, "Radio reflections from aurorae, III. The association with geomagnetic phenomena," *J. Atm. Terrestrial Phys. 11,* 237 (1957).

7. Chapman, J. H., B. C. Blevis, F. D. Green, H. V. Serson, and F. A. Camerson, "Preliminary data on radar returns from aurora at 488 Mc," Defence Research Telecommunications Establishment Project Report 44-2-1 (January 1958).

8. Chapman, S., "Radio echoes and magnetic storms," *Nature 122,* 768 (1928).

9. Chapman, S., "The geometry of radio reflections from aurorae," *J. Atm. Terrestrial Phys. 3,* 1 (1952).

10. Chapman, J. H., and B. W. Currie, "Radio noise from aurora," *J. Geophys. Research 58,* 363 (1953).

11. Dyce, R. B., Research Report EE 249, School of Electrical Engineering, Cornell University, Ithaca, N. Y., June 1955.

12. Forsyth, P. A., B. W. Currie, and F. E. Vawter, "Scattering of 56 Mc radio waves from the lower ionosphere," *Nature 171,* 352 (1953).

13. Forsyth, P. A., W. Petrie, and B. W. Currie, "Auroral radiation in the 3000 Mc region," *Nature 164,* 453 (1949).

14. Forsyth, P. A., W. Petrie, and B. W. Currie, "On the origin of ten centimeter radiation from the polar aurora," *Can. J. Phys. 28,* 324 (1950).

15. Forsyth, P. A., and E. L. Vogan, "The frequency dependence of radio reflections from aurora," *J. Atm. Terrestrial Phys. 10,* 215 (1957).

16. Fricker, S. J., S. M. Ingalls, M. L. Stone, and S. C. Wang, "UHF radar observations of aurora," *J. Geophys. Research 62,* 527 (1957).

17. Harang, L., "Investigations on the aurorae and the ionosphere at the Auroral Observatory, Tromso," *Trans. Edinburgh Meeting,* Int. Union of Geodesy and Geophysics, Association of Terrestrial Magnetism and Electricity, Bull. 10, 120 (1936).

18. Harang, L., "Scattering of radio waves from great virtual distances," *Terrestrial Magnetism and Atm. Electricity 50,* 287 (1945).

19. Harang, L., *The aurorae* (John Wiley, New York, 1951).

20. Harang, L., and B. Landmark, "Radio echoes observed during aurorae and terrestrial magnetic storms using 35 and 74 Mc waves simultaneously," *Nature 171,* 1017 (1953).

21. Hartz, T. R., G. C. Reid, and E. L. Vogan, "VLF auroral noise," *Can. J. Phys. 34,* 768 (1956).

22. Hartz, T. R., "Auroral radiation at 500 Mc," *Can. J. Phys. 36,* 677 (1958).

23. Hellgren, G., and J. Meos, "Localization of aurorae with 10 m high power radar technique using a rotating antenna," *Tellus 3,* 249 (1952).

24. Heppner, J. P., E. C. Byrne, and A. E. Belon, "The association of absorption and E_s ionization with aurora at high latitudes," *J. Geophys. Research 57,* 121 (1952).

25. Herlofson, N., "Interpretation of radio echoes from polar auroras," *Nature 160,* 867 (1947).

26. Hoag, J. B., and V. J. Andrew, "A study of short-time-multiple signals," *Proc. I. R. E. 16*, 1368 (1928).

27. Kaiser, T. R., "Radio investigations of aurorae and related phenomena," *Proc. of the Belfast Symposium on the Aurora and Airglow,* Vol. 5 of Special Supplements to the *J. Atm. Terrestrial Phys.,* 156 (Pergamon, London, 1955).

28. Leadabrand, R. L., L. Dolphin, and A. M. Peterson, Final Report, Contract AF 30(602)–1462, Stanford Research Institute, Menlo Park, California, 1957.

29. Little, C. G., and A. Maxwell, "Scintillations of radio stars during aurorae and magnetic storms," *J. Atm. Terrestrial Phys. 2*, 356 (1952).

30. Little, C. G., W. M. Rayton, and R. B. Roof, "Review of ionospheric effects at VHF and UHF," *Proc. I. R. E. 44*, 992 (1956).

31. Lovell, A. C. B., J. A. Clegg, and C. D. Ellyett, "Radio echoes from the aurora borealis," *Nature 160*, 372 (1947).

32. McNamara, A. G., "A continuously recording automatic auroral radar," *Can. J. Phys. 36*, 1 (1958).

33. Meek, J. H., "Correlation of magnetic, auroral, and ionospheric variations at Saskatoon," *J. Geophys. Research 58*, 445 (1953).

34. Meek, J. H., "East-west motion of aurorae," *Astrophys. J. 120*, 602 (1954).

35. Meinel, A. B., "Systematic auroral motions," *Astrophys. J. 122*, 206 (1955).

36. Moore, R. K., "Theory of radio scattering from the aurora," *Trans. I. R. E.,* PGAP-3, 217 (August 1952).

37. Nichols, B., "Drift motions of auroral ionization," *J. Atm. Terrestrial Phys. 11*, 292 (1957).

38. Nichols, B., "Drift motions of auroral ionization," Scientific Report No. 1, Contract AF 19(604)-1859, Geophysical Institute, University of Alaska, July 1957.

39. Nichols, B., "Auroral ionization and magnetic disturbances," *Proc. I. R. E. 47*, 245 (1959).

40. Peterson, A. M., L. A. Manning, V. R. Eshleman, and O. G. Villard, Jr., "Regularly-observable aspect-sensitive radio reflections from ionization aligned with the earth's magnetic field and located within the ionospheric layers at middle latitudes," Technical Report No. 93, Contract ONR 7D, Stanford University (1955).

41. Peterson, A. M., and R. L. Leadabrand, "Long-range radio echoes from auroral ionization," *J. Geophys. Research 59*, 306 (1954).

42. Størmer, C., "Short wave echoes and the aurora borealis," *Nature 122*, 681 (1928).

43. Størmer, C., *The polar aurora,* (Clarendon, Oxford, 1955).

44. Thayer, R. E., Auroral effects on television," *Proc. I. R. E. 41*, 161 (1953).

45. Thomas, L. H., "Short wave echoes and the aurora borealis," *Nature 123*, 166 (1929).

46. Tilton, E. P., "On the very highs," *QST, 28*, 41–43, 86 (1944).

47. van der Pol, B., "Short wave echoes and the aurora borealis," *Nature 122*, 878 (1928).

3

Ionospheric Scintillation of Radio Waves of Extraterrestrial Origin

ROBERT S. LAWRENCE

Scintillations of discrete radio sources involve apparent fluctuations in both amplitude and position. They are analogous to the twinkling and dancing of optical stars but, in the case of radio frequencies below about 500 Mc/s, they arise in the ionosphere rather than in the troposphere.

When the plane radio wave from a distant point source impinges upon the ionosphere, an irregular distortion is impressed upon the phase surface of the wave. The distortion is caused by irregular patches of varying refractive index, and it has no immediate effect upon the amplitude of the wave. As the distorted wave travels from the ionosphere toward the ground, various portions of the distorted wavefront interfere and so, by the time the wave reaches the ground, an intensity modulation is added to the phase distortion.

The intensity changes, called "amplitude scintillations," are much easier to measure than are the phase scintillations. The relation between the two depends upon the scale of the irregularities in the ionosphere and upon the distance between the observer and the diffracting layer. While the amplitude scintillations alone reveal some of the properties of the ionospheric layer, simultaneous knowledge of both amplitude and phase effects is considerably more useful.[1]

Phase scintillations are measured at Boulder, Colorado, by means of a phase-tracking and phase-sweeping interferometer.[2] This instrument can be used in its phase-tracking mode to provide a continuous record of phase scintillations, or it can be used with phase sweeping to provide a record from which amplitude and phase information can be extracted independently. Figure 1 shows such a recording of the radio source Cygnus A, taken at 53 Mc/s at an elevation angle of 49°

43

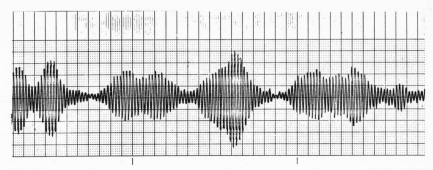

Fig. 1. A 12-min section of a phase-sweeping record of Cygnus A. The width of the envelope is proportional to the power received from the radio source.

on April 3, 1958. The record is 12 min long and shows violent amplitude scintillations. The phase information is contained in the details of the times of crossing of the center (zero) of the chart. This information is extracted from the record by means of a digital computer.

Figure 2 shows the computer output resulting from three of the phase-sweeping records. Each is divided into two parts giving separately simultaneous amplitude and phase variations. The duration of each record is 10 min. The phase variation has been magnified so as to occupy the entire vertical scale of the record, and the corresponding range is indicated in each case. The degrees referred to in this range are electrical degrees of phase difference in the signals from the two antennas of the interferometer. They can be converted into degrees in the sky by dividing by about 500.

The records in Fig. 2 exhibit the pronounced tendency of scintillations, in both amplitude and phase, to become larger as the elevation angle of the radio source decreases. This effect, primarily the result of the increased path length through the ionosphere, is complicated by the changing distance between the ionospheric irregularities and the observer. The fluctuation rates of the amplitude and phase variations appear to be quite different, and a careful study of the statistical relation between them promises to be most useful in measuring the distance to the irregularities. At the present time, it is not certain whether the diffraction responsible for any particular scintillation record occurs at a height of 100 or 400 km; there is some evidence that each height is effective at times, and some evidence for intermediate heights.

One of the features of the present experiment at Boulder is an attempt to clarify this question. Each day, soon after the rise of Cygnus A above the northeast horizon, ionospheric pulse soundings are made with a sweep-frequency, vertical-incidence sounder 300 km away in Nebraska. The sounder is so placed as to observe the very portion of the ionosphere which causes the scintillations recorded at Boulder. Comparison between these two methods of observing the ionosphere has verified the previously reported relation between scintillations and "spread-F," and strongly suggests an inverse relation between scintillations and "sporadic-E." If further investigation corroborates this relation, we will have definite knowledge that some of the scintillation activity arises at the 100-km height of the E-region.

All observers agree upon certain general properties of scintillations. For example, the intensity of the fluctuations increases with the zenith distance of the source and with the length of the radio waves. There is a pronounced nighttime maximum of occurrence of scintillations and an increase as the line of sight passes near the auroral zone. Principal outstanding problems are the determination of the height at which diffraction occurs and the explanation of the nighttime maximum.

Detailed discussions of radio-star scintillations and extensive bibliographies of the subject appear in the three references.[1-3]

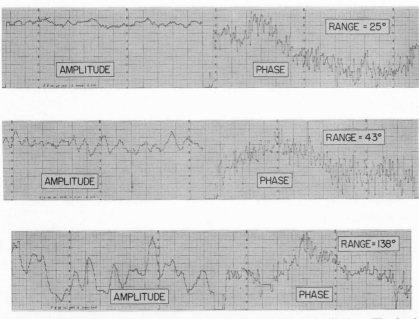

Fig. 2. Ten-minute records of simultaneous amplitude and phase scintillations. The depth of the amplitude fluctuation and the range of the phase variation both increase as the elevation angle above the horizon is reduced. Top: Cassiopeia A, 25 February 1958; 108 Mc/s, elevation, 71°. Center: Cygnus A, 28 February 1958; 108 Mc/s, elevation, 34°. Bottom: Cygnus A, 26 February 1958; 108 Mc/s, elevation, 15°.

Discussion

R. Jastrow, *Naval Research Laboratory:* Have you observed scintillations in your satellite tracking?

R. S. Lawrence: Ionospheric effects were observed upon our satellite records, but they do not look like radio-star scintillations. This is because the apparent angular velocity of the satellite is so great that the irregularities which produce scintillations would produce fluctuations of perhaps 10 c/s on a satellite signal. We have not yet measured such fluctuations in detail, although we know that they sometimes exist.

A. M. Peterson, *Stanford University:* In one case of satellite tracking with our 106-Mc/s radar at Stanford Research Institute, we observed a great increase in fading rate, when an aurora (I think it was the Feb. 10–11 aurora) occurred. The rate went up by a factor of 10 over the normal fading rate of the satellite echo.

S. Gruber, *AFCRC:* Have you performed any autocorrelation analysis of your data?

Lawrence: We have made autocorrelograms of some of the phase records, and from these have deduced power spectra. In general there are no predominant frequencies which occur regularly from one spectrum to another.

K. Toman, *Geophysics Research Directorate:* Do you distinguish between ordinary and extraordinary waves as the radio star makes a passage through the ionosphere? An extraterrestrial radio wave will, in entering the ionized atmosphere, be split into X and O, due to the magnetic field of the earth and perhaps due to the magnetic fields of local currents. Will their interference not also produce scintillations?

Lawrence: Our observations are made with plane-polarized receiving antennas, so that we get a combination of the two modes. However, in the case of a noise source the ordinary and extraordinary components are incoherent and there is no systematic interaction between them.

S. H. Autler, *M.I.T. Lincoln Laboratory:* Can you rule out either theoretically or experimentally the possibility of scintillation occurring in the dense non-ionized air of the troposphere?

Lawrence: The dependence of scintillations upon radio frequency shows conclusively that an ionized region is responsible. In the troposphere the refractivity is independent of frequency; in the ionosphere it varies inversely as the square of the frequency. Scintillations are observed to depend strongly upon frequency.

Gruber: But you do observe scintillations at higher frequencies than 50 Mc/s?

Lawrence: Yes, some of our records are taken at 108 Mc/s.

Gruber: At 200 Mc/s, or so, you might find that scintillations might be tropospheric, where the ionosphere does not have quite this effect. In experiments which I have performed, I noted a definite increase in fluctuation rate at times of visual aurora. Also, it might be noted that fluctuations of tropospheric origin might be observed at frequencies higher than those considered by Mr. Lawrence.

Lawrence: Yes.

REFERENCES

1. H. G. Booker, "The use of radio stars to study irregular refraction of radio waves in the ionosphere," *Proc. I. R. E. 46*, 298 (1958).
2. R. S. Lawrence, "An investigation of the perturbations imposed upon radio waves penetrating the ionosphere," *Proc. I. R. E. 46*, 315 (1958).
3. C. G. Little, W. M. Rayton, and R. B. Roof, "Review of ionospheric effects at VHF and UHF," *Proc. I. R. E. 44*, 992 (1956).

4

Meteor Scatter

VON R. ESHLEMAN

1. Introduction

The trails of electrons and ions created in the upper atmosphere by meteor particles will reflect and scatter radio waves. Every day more than 10^{11} such trails are formed by the many tons of interplanetary dust intercepted by the earth as it travels in its orbit. Because of the large number of meteor ionization trails and their scattering efficiency, they will propagate signals which can be used for long-range communications. In addition, meteor ionization trails may contribute to the propagation of terrestrial noise and unwanted signals into a receiving-antenna beam. Thus knowledge of the mechanism of scattering from meteor trails and information on the statistical variation of meteor echo characteristics are important in a description of meteor communication techniques and in a description of the radio noise spectrum.

Radio studies of meteors have led to new information on meteor astronomy and upper-atmosphere physics, and to important applications in radio communication. For example, it has been found that meteors are members of the solar system and that some (sporadic meteors) travel in independent orbits and others (shower meteors) in related orbits about the sun. Daytime meteor showers, which cannot be observed visually, have been discovered from the radio studies.[1] Recent high-sensitivity radar measurements show that there are vast numbers of ionizing meteors of masses less than 0.0001 those of the particles which produce the faintest light trails to be seen by eye.[2] The directions and speeds of the winds which blow in the 100-km height region have been studied by means of radio reflections from ionized meteor trails.[3-5] New information has also been obtained about the temperature and pressure in this region.[6] It is believed that

49

meteor ionization at times plays the dominant role in the propagation mode for ionospheric scatter communication.[7-14] The newly explored techniques of meteor-burst propagation promise to provide many new and more reliable channels for communication over distances up to about 2000 km.[15]

2. The Ionized Trail

The speed of a meteor particle relative to the earth's atmosphere lies between two limits. The lower limit, about 11 km/s, is the earth's escape velocity. An upper limit exists because meteors are members of the solar system and hence have elliptical orbits. The earth's orbital speed is about 30 km/s. A particle in a parabolic orbit (the limit between elliptical and hyperbolic orbits) near the earth would be going $2^{1/2}$ times as fast. A head-on collision between the earth and such a particle would result in a meteor speed of about 72 km/s relative to the atmosphere.

While a few large particles reach the ground, the vast majority of the trail-producing meteors are totally disintegrated in the region between about 85 and 115 km above the surface of the earth. At these heights the molecular mean free path (1 to 70 cm) is greater than the meteor particle diameter ($\ll 1$ cm). In the collisions between individual air molecules and the meteor, the air molecules may knock out a large number of meteor atoms from the parent particle. These meteor atoms then have speeds compounded from the original speed of the meteor and the much lower thermal velocity of their escape from the parent particle. When the free meteor atoms collide with the surrounding air molecules, many of the atoms, and perhaps some of the air constituents, are ionized. But after the first collision the meteor atoms (or, by now, ions) retain much of their high initial velocity and continue in nearly the same direction at greater than thermal speeds. In a recent study, Manning[16] has shown that the region of ionized gas formed near the meteor particle expands almost instantaneously by this process to about 14 ionic mean free paths, after which it continues to expand at a much slower rate by normal diffusion. Since an ionic mean free path is approximately one fifth of the molecular mean free path, this means that the initial radius of the ionized trail is between about 2 m at 115 km and 3 cm at 85 km.

As the meteor particle descends through progressively denser air, the number of ions (and electrons) formed per unit length increases slowly to a maximum and then decreases rapidly to zero as the particle is totally disintegrated.[17] The height of maximum ionization production is higher for greater speeds, greater zenith angles, and smaller particles. The mean height of maximum ionization for radar-detected sporadic meteors is about 93 km. The length of the straight column of ionization between points in a given trail, where the electron line density is half of the maximum, is approximately 2H sec ζ, where H is the scale height (about 7 km) and ζ is the zenith angle of the track of the meteor. The most likely (modal) length of the meteor trails detected by radio reflections is about 25 km.[18, 19]

After formation, the long, thin column of ionization expands by normal diffusion. The square of the radius r at time t is given by

$$r^2 = 4Dt + r_o{}^2, \tag{1}$$

where D is the diffusion coefficient (1 m^2/s at 85 km to 140 m^2/s at 115 km)[20] and r_o is the initial radius mentioned above. In this equation, the time t is measured from the time of passage of the meteor particle by the point along the trail that is being considered. For this reason, and because D and r_o vary rapidly with height, the radius of the trail at a given instant is different for different positions along the trail. Whether or not this is an important consideration in the determination of radio scattering from the trail depends upon several factors, as discussed below.

3. Mechanism of Radio Scattering

Ideally, in a theoretical development, it would be desirable to have an exact model of the phenomenon under study. Nearly always in practice, however, it is necessary to deal with exact solutions for an approximate model or approximate solutions for a more nearly exact model. Often the only feasible procedure is to obtain an approximate solution for an approximate model.

Approximate solutions are often very valuable even when exact solutions have been obtained for the same model. This is especially true when the exact solutions require numerical methods, while the approximate solutions yield explicit expressions showing the contributions of each of the variables to the results. Under this condition the approximate solutions can be used to investigate the effects of all possible values of the various parameters, while the exact solutions, made for several discrete combinations of the variables, serve as valuable references from which the limitations and reliability of the approximate results can be evaluated.

In the past 10 years, more than 20 authors have written more than 30 theoretical papers on the mechanism of radio scattering from meteor ionization trails. The treatments in all of these papers can be classed as various grades of exact and approximate solutions for approximate models of the meteor ionization trails. Perhaps the most ambitious computational efforts were those of Keitel[21] and of Loewenthal.[22] Both authors used high-speed electronic digital computers for solving complex integral equations. Keitel considered right circular cylinders of ionization, and found the internal cylindrical wave modes to be matched to the external radio waves. In general, the external radio waves are expressed as an infinite sum of cylindrical waves, and each internal cylindrical wave is found from an infinite series. Thus, finding the radio echo from a trail involves an infinite sum, each term of which is itself an infinite sum. Little wonder that only a few exact solutions of this type have been obtained. While the circular cylinder is a good approximation to the actual trail at relatively long radio wavelengths, a different model is required for wavelengths shorter than about 3 m. Loewenthal considered a more accurate model for short wavelengths, but could treat it only by assuming first-order electron scattering (that is, a relatively low density of electrons). Even then, the solution requires the integration of Fresnel integrals with complex limits, a job suitable only for high-speed computers.

While it is not feasible to apply the computer techniques to find the scattering characteristics of meteor trails of many different sizes, shapes, and densities, the few computations that have been made show that some of the more approximate analytical methods which have been used are sufficiently accurate for most purposes. Therefore the results of the approximate analyses will be reviewed here, with mention of their limitations when appropriate.

Different assumptions and techniques are used in the approximate analyses according to the density of ionization, the radio wavelength, and the geometry of the propagation path. The density of importance in meteor trail scattering is not the volume density (the number of electrons per cubic meter) but the line density (the total number of electrons per meter of length along the trail). The division between high and low line density is about 10^{14} electrons/m, the approximate density produced by a faint visual (fifth-magnitude) meteor. For low densities, the approximate analyses consist of summing the scattered waves from the individual electrons, assuming that the incident radio wave passes through the ionized region unchanged. For high densities, ray tracing through the ionized region is used. The division in wavelength (long- and short-wavelength approximations) comes about, as mentioned above, because of the need for a more accurate trail model for the shorter wavelengths than the right circular cylinder that is a convenient model for longer wavelengths. The geometry of the propagation path controls some of the echo characteristics, so it is necessary to consider whether or not the transmitter and receiver are at the same location (radar case) or at widely separated locations (communication case, also referred to as "oblique propagation" or "forward scattering"). The formulas of echo strength and duration for the radar case will be presented, followed by a description of the required changes for application to oblique geometry.

The specular echo characteristics for the following four sets of parameters will be considered separately: low density, long wavelength; low density, short wavelength; high density, long wavelength; and high density, short wavelength. In a final section the theory for nonspecular echoes is presented.

A. Low Density, Long Wavelength. It is assumed here that the meteor trail is formed initially as an infinitely long right circular cylinder with a distribution in radius r of electron volume density given by $(q/\pi r_o^2) \exp(-r^2/r_o^2)$, where r_o is the initial radius discussed above and q is the number of electrons per meter of length along the trail. Subsequently, the ionization expands by normal diffusion so that the volume density at time t for any radius and position along the length of the trail is

$$\frac{q}{\pi r_o^2 + 4\pi Dt} \exp\left(-\frac{r^2}{r_o^2 + 4Dt}\right), \qquad (2)$$

where D is the diffusion coefficient.

Almost all of the scattered energy from this trail will come from the principal Fresnel zone, centered at the point of tangency of the meteor trail axis and one of a set of confocal prolate spheroids (spheres for the radar case) the foci of which are at the transmitter and receiver. Referring to this point as the position of the meteor trail, the ratio of the received power P_A to transmitted power P_T for the radar case is[23-25]

$$\frac{P_A}{P_T} = IEq^2 \frac{R\lambda}{2} \exp\left(-\frac{32\pi^2 Dt}{\lambda^2}\right), \qquad (3)$$

where I, the ratio of received to transmitted power for one electron, is $(GA\ r_e^2)/(4\pi\ R^4)$; G is the power gain over an isotropic source of the transmitting antenna in the direction of the meteor trail; A (m²) is the aperture of the receiving antenna for the direction and polarization of the scattered wave; R (m) is the radar range to the meteor trail; $r_e = \mu_o e^2/4\pi m = 2.8178 \times 10^{-15}$ m is the classical electron radius, where μ_o is the permeability of free space and e and m are the electron's charge and mass; $E = \exp[-(8\pi^2 r_o^2/\lambda^2)]$ is the attenuation of the echo due to the initial radius; and λ (m) is the radio wavelength.

The echo is maximum at $t = 0$, after which the intensity decays exponentially with a time constant τ_A given by [24, 25]

$$\tau_A = \frac{\lambda^2}{32\pi^2 D}.$$

(4)

This time constant, the time at which the intensity is 36.8 percent of its maximum value, is used as a measure of echo duration. The radar geometry and echo shape are illustrated in Fig. 1.

The subscript A is used on P_A and τ_A to indicate that the intensity and duration formulas are for the conditions being considered in section A, namely, low densities and long wavelengths. Subscripts B, C, D, and E will be used in later expressions for echo characteristics

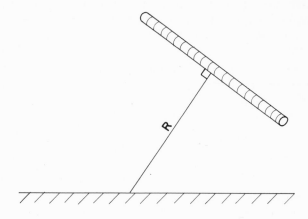

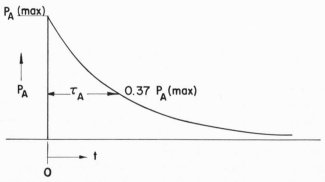

Fig. 1. The trail model, radar geometry, and echo shape for low densities and long wavelengths.

under the other conditions of density, wavelength, and geometry which are considered in the correspondingly lettered sections.

For oblique propagation the above formulas can be changed to give forward-scattered echo intensity and duration by:[9, 26] (1) replacing all λ's by $\lambda \sec \phi$; (2) replacing all R's, except in the expression for I, by $M(1 - \sin^2\phi \cos^2\beta)^{-1}$; and (3) changing I to read $GAr_e^2 \sin^2\alpha / 4\pi R_1^2 R_2^2$.

In these last expressions, R_1, R_2 (m) are the ranges from transmitter and receiver to the trail; 2ϕ is the included angle between R_1 and R_2; $M = 2R_1 R_2 \cos \phi (R_1 + R_2)^{-1}$ (m) is the length of the bisector of the angle 2ϕ in the triangle formed by R_1, R_2, and a straight base line; β is the angle between the trail axis and the plane of R_1, R_2; and α is the angle between the electric vector at the meteor trail and R_2.

Of particular importance in these formulas is the prediction that meteor echoes endure $\sec^2\phi$ times longer when observed over an oblique path than when detected by radar.[9] (While a large increase in duration for oblique propagation has been observed, it may not be as high as predicted for very long paths, where $\sec^2\phi$ is as high as 25.[26]) Also, the effect of the initial radius attenuation (E) is reduced exponentially for meteor detection over an oblique path as compared with radar detection.

More exact computations show a resonance effect in the echo when the incident wave is polarized perpendicular to the axis of the trail.[24-28] This resonance occurs only once in the lifetime of an echo, and at most doubles the echo amplitude and quadruples the echo intensity. If the motion of the meteor particle is included in the model, small Fresnel fluctuations in the time variation of echo intensity are obtained. These fluctuations are important for experimental determinations of the velocities of the meteor particles,[29-31] though they are of little significance in considerations of total signal energies.

B. Low Density, Short Wavelength. At the long wavelengths the meteor particle traverses the principal Fresnel zone, of length $(2\lambda R)^{1/2}$ m, in a time which is short compared to the time required for the ionization to expand to a size where the scattering per unit length is severely attenuated. (The traverse time is proportional to $\lambda^{1/2}$, while the expansion time is proportional to λ^2.) For the shorter wavelengths, however, the traverse time may become comparable to or greater than the expansion time. Then it is necessary to use a model of the trail which includes the change in radial size of the column with position along the trail axis.[22, 32-36] For the short wavelengths, therefore, the trail model is a paraboloid of revolution, with the pointed end truncated and replaced by a hemisphere corresponding to the initial radius.

Only a short section of the paraboloid of revolution will be small enough in radius to contribute strongly to the echo. Since the "local" duration is proportional to $\lambda^2 D^{-1}$, the length of the effective region is proportional to $\lambda^2 D^{-1} v$, where v is the meteor velocity. As long as this length is less than the length of a Fresnel zone, the q electrons per unit length along this effective length will scatter coherently. Thus the maximum echo intensity will be proportional to $q^2 \lambda^4 v^2 D^{-2}$. The echo duration will correspond to the time required for the meteor particle to travel between the two Fresnel zones on either side of the principal zone the lengths of which are comparable to the effective scattering length. From this consideration it can be shown that the echo duration at low densities and short wavelengths should be proportional to $DR\lambda^{-1} v^{-2}$.

These qualitative considerations are verified from a more careful analysis. The theoretical echo intensity, including the time variation, has been found[33] to be

$$\frac{P_B}{P_T} = IEq^2 \frac{(v\lambda^2/16\pi^2 D)^2}{1 + (v^2\lambda/4\pi DR)^2 t^2}.$$ (5)

The trail model and echo shape are shown in Fig. 2. The maximum value of P_B occurs when the meteor particle is at the center of the principal Fresnel zone ($t = 0$), and the echo intensity is symmetrical about this time. The duration τ_B is taken as the time during which the echo intensity exceeds $(1 + \pi^2/4)^{-1} = 28.8$ percent of its maximum value, so that

$$\tau_B = \frac{4\pi^2 DR}{\lambda v^2}.$$ (6)

The definition of duration used here and in the long-wavelength formula (4) corresponds to the duration of a rectangular pulse whose intensity is equal to the maximum echo intensity and whose energy is equal to the echo energy.

Note that the maximum value of (3) is equal to the maximum value of (5), and (4) equals (6), when

$$\lambda = \left(\frac{128\pi^4 D^2 R}{v^2}\right)^{\frac{1}{4}}.$$ (7)

Thus this value of λ marks the division between long and short wavelengths. For average values of the various parameters, the transitional wavelength is near 3 m for the radar geometry, while for long oblique paths it is near 0.5 m for across-the-path trails ($\beta = 90°$) and 1.5 m for along-the-path trails ($\beta = 0°$). The extreme values of the transitional wavelength are about 0.1 m and 100 m.

Of course the echo characteristics do not change abruptly between the long- and short-wavelength expressions at the transitional wave-length. Approximate formulas for maximum echo intensity and duration have been developed[32, 33] which apply to both long and short wavelengths, and which bridge the transition region smoothly. These are

$$\frac{(P_{A, B})_{max}}{P_T} = IEq^2 \left(\frac{v\lambda^2}{16\pi^2 D}\right)^2 T^2 \tag{8}$$

and

$$\tau_{A, B} = \frac{4\pi^2 DR}{\lambda T^2 v^2}, \tag{9}$$

where $T = 1 - e^{-2C} = 2e^{-C} \sinh C$, and $C^2 = 32\pi^4 D^2 R\lambda^{-3}v^{-2}$.

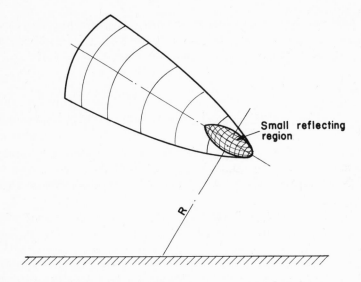

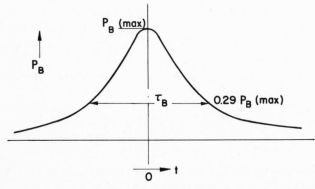

Fig. 2. The trail model, radar geometry, and echo shape for low densities and short wavelengths.

As before, more precise numerical analyses for the short wavelengths show small effects due to polarization resonance and Fresnel diffraction.[22, 35]

While the echo intensity for the short wavelengths is much less than for long wavelengths, the echo durations are longer. It is interesting to note that the echo energies for low-density trails at both long and short wavelengths are given by the same expression:[33]

$$P_T I E q^2 \frac{R\lambda^3}{64\pi^2 D} \text{ j.} \tag{10}$$

All of the above expressions can be applied to the oblique propagation geometry by making the same changes that are listed in the previous section.

The heights of the meteor trails strongly affect the echo energies through large variations of r_o and D. If the values of these parameters at 95 km are used for reference, the echo energy at a greater height h is reduced by a factor ρ_{AB}, assuming other parameters constant, given by

$$\rho_{AB} = \frac{D(h)}{D(95 \text{ km})} \quad \exp \left\{ \frac{8\pi^2}{\lambda^2} [r_o{}^2(h) - r_o{}^2(95 \text{ km})] \right\} \tag{11}$$

Curves for $\rho_{AB} = 10$, 10^2, and 10^3 in height-wavelength coordinates are presented in Fig. 3. The dashed line is the locus where the magnitude of the effects of D and r_o are equal. Below this line, D is more important than r_o in setting a "ceiling" above which it is difficult to obtain meteor echoes, while above the dashed line the effects of r_o predominate. Note that the "diffusion ceiling" is easier to penetrate with increased sensitivity than is the much more solid "initial-radius ceiling."[33, 36] Since λ is changed to $\lambda \sec \phi$ for forward scatter, the ceilings are higher for oblique geometry than for the radar geometry.

C. *High Density, Long Wavelength.* When the line density is sufficiently high, the assumption that the incident wave passes through the ionized region is no longer tenable. It is then convenient to assume that total reflection occurs at the surface where the ionization density corresponds to an equivalent zero dielectric constant. Then, using the same trail model as in Sec. A above (except for the greater density), the received echo intensity for the radar geometry is[25, 32, 37]

$$\frac{P_C}{P_T} = \frac{I q^{1/2} \lambda R}{4\pi r_e{}^{3/2}} \left[\left(\frac{t + t_o}{\tau_z} \right) \ln \left(\frac{\tau_z}{t + t_o} \right) \right]^{1/2}, \tag{12}$$

where $t_o = r_o{}^2/4D$, a diffusion time equivalent of the initial radius, and $\tau_z = \lambda^2 q r_e / 4\pi^2 D$, the duration of the high-density echo for zero initial radius. From (12) the total echo duration τ_C is

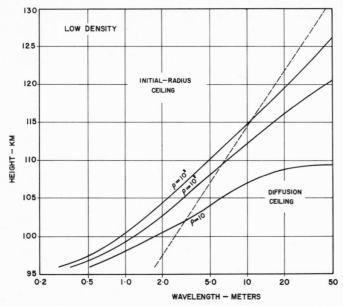

Fig. 3. Curves of constant degradation of echo energy for low-density trails due to the diffusion coefficient D and the initial radius r_o. The ratio of echo energy with D and r_o at their 95 km values to the echo energy at the indicated height is given by ρ. The dashed line is the locus where the effects of D and r_o are equal.

$$\tau_C = \tau_z - t_o, \tag{13}$$

and the received echo intensity is maximum at time $\tau_z e^{-1} - t_o$ with the value

$$\frac{(P_C)_{\max}}{P_T} = \frac{Iq^{1/2}\lambda R}{4\pi r_e^{3/2}} e^{-1/2}. \tag{14}$$

The trail model and echo shape are illustrated in Fig. 4. In (14) it is assumed that the initial radius is smaller than the maximum radius that would be obtained in the absence of an initial value. If $\tau_z > t_o > \tau_z e^{-1}$, the maximum echo intensity occurs at zero time and has the value obtained by placing $t = 0$ in (12). The total echo duration is still given by (13). If $t_o > \tau_z$, the principal Fresnel zone is not effective and reflection occurs only at the rapidly expanding coma near the meteor particle (see Sec. E).

Note that if we discount for the moment the effect of the initial radius, (3) is equal to (14) and (4) is equal to (13) for the radar case when q is approximately 10^{14} electrons/m. Thus this value of line density is used as the transition density between low- and high-density approximations.

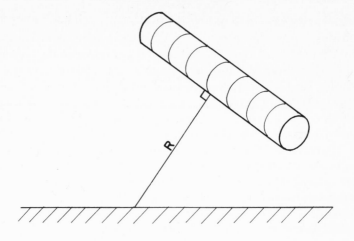

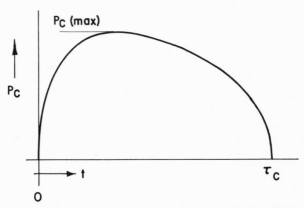

Fig. 4. The trail model, radar geometry, and echo shape for high densities and long wavelengths, assuming $r_o \cong 0$.

For the low-density trails, echo intensity is proportional to q^2 and echo duration is independent of q, while, for high densities, intensity is proportional to $q^{1/2}$ and duration is directly proportional to q.

It is tempting to expand the preceding radar formulas to include general oblique geometry by making the same changes as have been found applicable in the low-density case—by changing λ to $\lambda \sec \phi$ and R to an equivalent distance parameter which includes the effect of trail orientation relative to the plane of propagation. If this could be done, echo durations would be proportional to $\lambda^2 \sec^2 \phi$, just as for the low-density exponential echoes.[38] However, Manning[39] has shown from computations of ray paths through the cloud that echo duration is not exactly proportional to $\sec^m\phi$ for any value of m, though the durations can be closely approximated by a $\sec^2\phi$ variation for along-the-path trails and a $\sec^{0.3}\phi$ variation for across-the-path trails. For a random distribution of meteor trail orientations, the average value of m is about 0.85. In every case the duration is proportional to λ^2, assuming the initial-radius effect to be small. The ray-tracing technique also indicates that the maximum echo intensity is somewhat less than that given by (14), although this variation is not as important as the duration characteristics summarized here. Exact analysis of the high-density, long-wavelength model is extremely difficult because of the slow convergence of the infinite series which are involved. The ray-tracing solution compares well with the one exact analysis that has been made.[21, 39]

Most long-wavelength echoes from high-density trails endure long enough for the trail to become distorted by wind shears in the upper atmosphere, so that Fresnel-zone reflections may occur at from several to many points along the length of the trail. The scattered waves from these "glints" interfere at the receiver and cause fading of the echo. While it is agreed that the large-scale bending of the trail produces the fading, there is some controversy in the literature concerning the local reflection and expansion process. In one view,[40-42] the trail expands by normal diffusion while its axis is contorted into a sinuous shape by large-scale effects, so that each of several principal Fresnel zones along the trail reflects radio energy according to the foregoing formulas. But Booker and Cohen[43] have suggested that local expansion and scattering properties are controlled by small-scale (order of 1 m) turbulence which renders each portion of the trail rough in about 0.5 sec. If the trails are rendered rough in this manner, high-density trails would produce echoes within a half-second of their formation regardless of the orientation of the trail. Manning[40] and Manning and Eshleman[42] have pointed out, however, that the combined visual and radar results of McKinley and Millman[44] show that aspect sensitivity persists for an average of 10 sec after formation. That is, a high-density trail oriented along a radial line from a radar site does not produce an echo until after about 10 sec, when the large-scale wind shears have bent parts of the trail by 90° so that Fresnel-zone reflections are possible. It is also pointed out by Manning that the existence of small-scale turbulence is not needed for the explanation of any of the experimental results, and indeed is inconsistent with many experimental results. Thus the model for high-density trails at long wavelengths is one which gives Fresnel reflections according to the preceding formulas at each portion along its length that is contorted into a favorable orientation by large-scale (order of 1 km) inhomogeneities in the upper atmosphere.

D. *High Density, Short Wavelength.* As the meteor trails expand radially, high-density trails continue to scatter strongly at larger trail diameters than for the low-density trails. Nevertheless, the combined action of the initial radius and rapid diffusion at short wavelengths can make the effective scattering length shorter than the principal Fresnel zone. Since the duration of the reflection at a given point along the trail axis is τ_C sec, the length of the scatterer is $\tau_C v$ m. The maximum echo strength per unit length can be approximated by $(P_C)_{max}/(\lambda R/2)$, so that

$$(P_D)_{max} = \frac{(2P_C)_{max}}{\lambda R} (\tau_C v)^2. \tag{15}$$

If the initial-radius effect is neglected in τ_C, expression (15) varies in the same way with the parameters λ, v, and D as the expression obtained by Hawkins and Winter[45] for the maximum echo strength for high-density meteor trails at short wavelengths. Their formula corresponds to $(P_D)_{max}$ equal to $(P_B)_{max}$ with r_o set equal to zero. Note that in (15) $(P_D)_{max}$ varies as $q^{5/2}$, while Hawkins and Winter indicate that the echo intensity varies as q^2. The two expressions are equal if r_o is set equal to 0 when q equals 10^{14} electrons/m, with (15) larger for all greater line densities. It appears that the different line-density dependence is due to the use by Hawkins and Winter of a parabolic radial distribution of electron density as compared to the Gaussian model assumed here. A q^2 dependence results in both cases from the effective scattering length. Only in the model used for (15), however, is the effect of the dependence of the diameter of the target on q reflected in an additional dependence of echo intensity on line density. The method used in deriving (15) also includes the effect of the initial radius (in τ_C), and, as shown below, leads to an expression for echo duration.

As in Sec. B, echo duration can be approximated by the time required for the effective scattering region to travel between the two Fresnel zones whose lengths are comparable to the scattering length. The time of travel between the two portions of the nth Fresnel zone is $(2n\lambda R)^{1/2}/v$. The length of the nth Fresnel zone is $[n^{1/2} - (n-1)^{1/2}]$ $(\lambda R/2)^{1/2}$ or approximately $(\lambda R/8n)^{1/2}$. Setting this length equal to the scattering length $\tau_C v$, solving for n, and using this n in the expression for the travel time, we find for the echo duration

$$\tau_D = \frac{\lambda R}{2\tau_C v^2}. \tag{16}$$

The echo shape is probably similar to that shown in Fig. 2.

If the effects of the initial radius r_o are small, (5) equals (15) and (6) equals (16) when the line density is approximately 10^{14} electrons/m. Thus this marks the division between high and low densities for short wavelengths just as for long wavelengths. The transition between long and short wavelengths for high densities depends upon D, R, and v in the same manner as shown for low densities in (7), but, in addition, depends upon $q^{-2/3}$. Thus for the radar geometry the average transitional wavelength is progressively less than about 3 m for increasing densities above 10^{14} electrons/m.

The modifications which must be made in the preceding formulas for application to forward scattering can be determined from the geometric changes plus a consideration of Manning's results on ray tracing.[39] At short wavelengths, the echo intensities and durations for high-density trails near the path mid-point ($R_1 = R_2 = R$) depend upon sec ϕ as follows: for along-the-path trails, intensity proportional to $\sec^4\phi$ and duration independent of ϕ; for across-the-path trails, intensity proportional to $\sec^{0.6}\phi$ and duration to $\sec^{-0.3}\phi$.

Taking the echo energy as $(P_D)_{max}\tau_D$, which is equal to $(P_C)_{max}\tau_C$, the echo energy for high densities is reduced at height h by diffusion and the initial radius from its value at 95 km by the factor

$$\rho_{CD} = \frac{D(h)}{D(95 \text{ km})} \cdot \frac{\lambda^2 q r_e - \pi r_o{}^2(95 \text{ km})}{\lambda^2 q r_e - \pi r_o{}^2(h)}. \tag{17}$$

This factor is plotted for $q r_e = 1$ and $q r_e = 10$ in height-wavelength coordinates in Fig. 5.

E. Nonspecular Echoes and Doppler Frequencies. In the previous sections only the strong echoes obtained when the meteor particle passes through the principal and adjacent Fresnel zones were considered. But if the particle is completely vaporized before reaching this region, or if it starts producing ionization after having passed these zones, weak echoes can still be obtained. These echoes result from scattering at the ion-density discontinuity near the particle at the head of the column. These echoes are called "head echoes," meteor "whistles," or nonspecular echoes. For low densities at both long and short wavelengths, the nonspecular echo strength is approximately

$$\frac{P_E}{P_T} = I E q^2 \left(\frac{\lambda R}{4\pi v t}\right)^2. \tag{18}$$

In this expression, R is the range from the radar site to the position of the meteor particle, and it is assumed that this position is beyond the first several Fresnel zones. The time t is measured relative to the time of crossing the principal zone. Since the nonspecular echo moves with

64

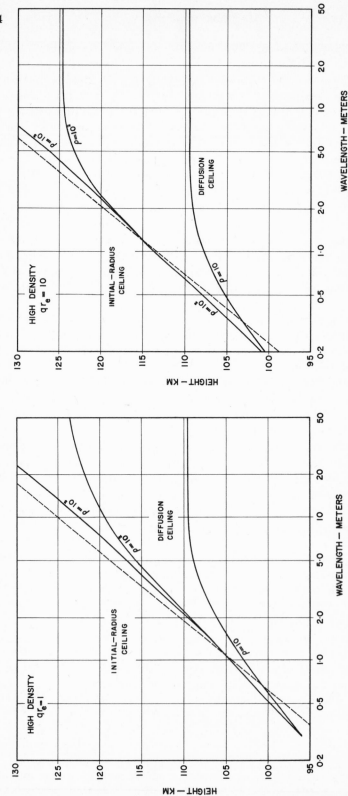

Fig. 5. Curves of constant degradation of echo energy for high-density trails due to the diffusion coefficient D and the initial radius r_o. The ratio of echo energy with D and r_o at their 95 km values to the echo energy at the indicated height is given by ρ. The dashed line is the locus where the effects of D and r_o are equal. (l.) $qr_e = 1$, $q = 3.56 \times 10^{14}$ electrons/meter. (r.) $qr_e = 10$, $q = 3.56 \times 10^{15}$ electrons/meter.

the meteor particle, the echo duration τ_E is L/v sec, where L is the length of the trail in meters.

The foregoing treatment for low-density nonspecular echoes does *not* apply to the rare head echoes recorded by McKinley,[46] which are obviously from high-density trails. Equation (18) should apply, on the other hand, to the numerous meteor whistles which can be heard with sensitive receiving equipment monitoring a distant VHF high-powered transmitter.

The frequency of the received signal from nonspecular scattering is shifted owing to the motion of the particle by the Doppler frequency f_E given by

$$f_E = -\frac{2v^2 t}{\lambda R}. \tag{19}$$

Winds also impart a frequency shift to the echoes received by specular reflections. The rms "body Doppler" shift due to winds is about 20 c/s at a radar frequency of 100 Mc/s, and is linearly related to the frequency.

Equations (18) and (19) may be extended to apply to the oblique geometry by the same procedure as that used in Secs. A and B.

A satisfactory theory of nonspecular echoes from high-density trails has not yet been evolved. It may be that the initial radius in a very-high-density trail is made even larger than suggested above by the action of the large amount of heat evolved from the meteor. In this case only the region near the particle would be dense enough to scatter radio waves during the rapid transient diffusion process, so that a strong head echo could be obtained even though no long ionization wake were left which could cause a specular type of echo ($t_o > \tau_z$). This is the kind of model needed to explain the measurements of the unusual head echoes made by McKinley.

4. Meteor Echo Statistics

If it is assumed that the light and ionization created by meteor particles are directly proportional to their initial mass, visual[47] and radar[32, 48] studies indicate that the number of sporadic meteors of mass greater than m is inversely proportional to m. In most meteor showers, the proportion of larger meteors is greater than for the sporadic meteors.[48] Sporadic meteors are much more important than recognized showers in meteor scatter propagation, so the mass distribution mentioned above will be used to find the number-intensity and number-duration distributions of specular meteor echoes. The diurnal and seasonal variations in the number and directions (radiants) of the meteors will also be discussed.

A. Number-Intensity Distribution. For the long wavelengths, specular-echo intensity is proportional to q^2 for low densities and to $q^{1/2}$ for high densities. Since the number of sporadic meteors of mass greater than m is approximately inversely proportional to m, the number of trails with line density greater than q is inversely proportional to q. Thus $(N_P)_{1.w.} \sim P^{-0.5}$ for low densities and $(N_P)_{1.w.} \sim P^{-2.0}$ for high densities. This variation is plotted in Fig. 6. Of course the transition between the two curves is not sharp, and more detailed analyses show a gradual transition between the two lines shown in the figure. The predicted number-intensity distribution agrees well with experiment.[25, 49]

For short wavelengths, since echo intensity is proportional to q^2 at low densities and $q^{5/2}$ at high densities, $(N_P)_{s.w.} \sim P^{-0.5}$ for low densities and $(N_P)_{s.w.} \sim P^{-0.4}$ for high densities. These variations are shown in Fig. 7. This predicted behavior has not yet been checked experimentally.

B. Number-Duration Distribution. For long wavelengths the time constant of the low-density echoes τ_A is independent of line density. However, the duration above a threshold (noise) level depends on q according to $\tau_A \ln(q/q_o)^2$, where q_o is the line density corresponding to an echo whose peak intensity is equal to the threshold level. For high densities, τ_B is a total duration which is little affected by a changing threshold, and its value is proportional to q for small initial radii. Thus the number of trails of duration greater than τ is $(N_\tau)_{1.w.} \sim \exp(-\tau/2\tau_A)$ for low densities and $(N_\tau)_{1.w.} \sim \tau^{-1}$ for high densities.

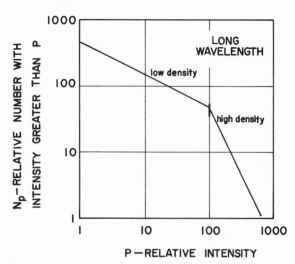

Fig. 6. Number-intensity distribution of meteor echoes at long wavelengths.

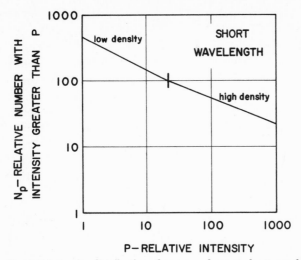

Fig. 7. Number-intensity distribution of meteor echoes at short wavelengths.

These curves are presented in Fig. 8 for an arbitrary ordinate relation between the two curves. The shape of this combination of curves appears to correspond to experimental measurements.[26]

For short wavelengths the low-density echo durations to a given fraction of peak intensity are independent of q, but the durations above a threshold corresponding as before to q_o are proportional to $[(q/q_o)^2 - 1]^{1/2}$. For high densities and small initial radii the echo durations are inversely proportional to q. Thus the few high-density trails have durations which overlap in value the durations of the more numerous low-density trails. This overlap will only slightly affect the

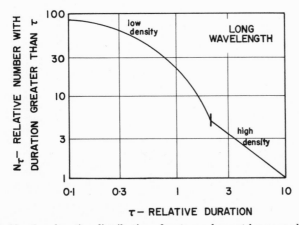

Fig. 8. Number-duration distribution of meteor echoes at long wavelengths.

number-duration curve, so that the number of echoes of duration greater than τ for short wavelengths and including both high and low densities can be taken as $(N_\tau)_{s.w.} \sim (\tau^2 + 1)^{-1/2}$ up to the maximum duration (which occurs for trails of the transitional density), after which $(N_\tau)_{s.w.} = 0$. The curve for the foregoing number-duration distribution with an arbitrary maximum duration is plotted in Fig. 9. This predicted behavior has not yet been verified experimentally.

C. *Rate and Radiant Variations.* The most important single factor controlling the number (rate) and directions of arrival (radiants) of meteors at a particular geographic location is the geometric relation between this location and the apex of the earth's way (the direction of the velocity vector of the earth in its orbit). As the earth sweeps through the cloud of interplanetary dust at 30 km/s, more meteoric impacts occur on the morning side of the earth (the leading side) than on the evening (trailing) side. The ratio of the maximum to minimum rate usually varies between about 3 and 6, but may at times be as great as 20. Because of the tilt of the earth's axis from the poles of the ecliptic, the amount of the diurnal variation decreases with increasing latitude, the rate becoming constant throughout the day at the poles. Similarly, the magnitude of the seasonal variation of rate (being maximum at the autumnal equinox and minimum at the vernal equinox) increases with increasing latitude of the observer. In addition there appears to be an increase in meteor rate during June and July (in both Northern and Southern Hemispheres) because of a nonuniformity in the number of meteor orbits crossing various parts of the earth's orbit.[50]

Since strong meteor echoes are obtained only when the ionized trail is positioned so as to have a principal Fresnel zone, the changing

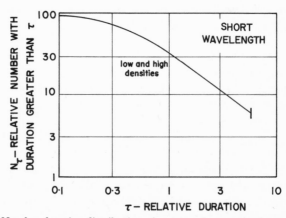

Fig. 9. Number-duration distribution of meteor echoes at short wavelengths.

rate and radiant distributions of incoming particles both affect the rate of detection of meteor echoes.[9, 51-53] For maximum detection rate in mid-northern latitudes, the antenna beam of a radar system should be pointed north at 0600, east at 1200, south at 1800, and west at 2400 local time. These same considerations (see Eshleman and Mlodnosky[53]), based on the apex concentration of meteors, indicate that for maximum meteor echo detection rate over oblique paths the antenna beams at transmitter and receiver should be directed not along the great-circle path, but to one side or the other of the direct path. For an east-west circuit in mid-northern latitudes, the antennas should be directed north of the great-circle bearing for the hours centered on 0600 and south of the direct bearing for the hours centered on 1800 local time. For a north-south path, the favored deviation from the direct bearing is toward the west at night and toward the east during the day. For radars and oblique circuits in mid-southern latitudes, interchange north and south in the previous statements.

The foregoing predictions of the gross directional features of meteor echo detection have been confirmed experimentally. From radar measurements taken at Stanford University the direction from which the maximum echo rate can be obtained varies with time of day as indicated above. The approximate ratio of the morning maximum to evening minimum echo rate depends upon antenna beam direction as: (1) north beam, 6; (2) east beam, 3; (3) south beam, 2; and (4) west beam, 3. Measurements of the direction of arrival of meteor echoes propagated over 1000-km north-south and east-west paths[53, 54] show that the side of the highest rate changes essentially as predicted above, with the off-path angle for maximum rate varying from a few degrees to more than 20°. The number of echoes received from one side of the path may be as much as 10 times the number of echoes from the other side of the path.

These predictions are for the so-called "sporadic meteors." If the rate of detection of echoes is high, most of the recognized meteor showers have little effect on meteor scatter characteristics. That is, most of the signals during shower periods are due to small sporadic meteors and not to the larger shower meteors. However, sudden changes in the directional properties of oblique meteor scatter have been noted.[53, 54] These strong fluctuations in the meteor radiant distribution must be due to heretofore undetected showers of small particles (around 10th magnitude). These effects may last only about an hour, indicating that the particles are tightly bunched in their orbit.

Recent measurements at Stanford University[55] indicate that most of the very small radio meteors (15th magnitude) do not occur completely at random, but may be disposed into a very large number of shower orbits. This conclusion is based on the fact that the day-to-fluctuations in echo rate are much greater than would be obtained from a population of independent particles. The small 15th-magnitude meteors are believed to be responsible for much of the signal propagated in ionospheric scatter communications.[9, 56, 57] The larger 10th-magnitude particles are important in meteor burst propagation. More intensive future study may well show that most of the "sporadic meteors" are really members of a large number of showers.

The main directional features of oblique meteor scatter propagation are not dependent upon whether the small meteors are sporadic or in many groups. However, finer scale variations in the directional properties of meteor scatter do depend upon a more exact description of the meteor radiants than is available at this time.

5. *Signal Characteristics*

The preceding discussion covers two areas: (1) the manner in which echoes from individual meteor trails vary with parameters of the ionized trail, the propagation path, and the radio system; and (2) the distributions of echo intensities, echo durations, and trail numbers and orientations. These results can now be combined to describe the signal which is propagated by many meteor trails from a distant transmitter. The meteor scatter signal can be described in terms of the number of echoes per unit time, the average signal power, the median signal power, or the fraction of the time (duty cycle) that meteor echoes are present above a threshold level.

Table 1 lists the foregoing characteristics of the total signal propagated by many meteor trails, together with the single echo characteristics and the intensity and duration distributions. The list shows how the indicated quantities vary with transmitter power P_T, receiver power threshold L, antenna gain G, trail line density q, diffusion coefficient D, meteor particle velocity v, range R, wavelength λ, forward-scattering angle 2ϕ, echo intensity P, and echo duration τ. The effect of the initial radius r_o is not included. (It could be included, however, in all of the low-density formulas in Table 1 by putting $E = \exp(-8\pi^2 r_o^2/\lambda^2\sec^2\phi)$ with P_T wherever it occurs. Allowing for r_o in the high-density formulas would be more involved, although it would follow from the formulas in Secs. 3C and D.) For the oblique path, λ is changed to $\lambda \sec \phi$ in intensity and duration formulas for low densities, and to $\lambda \sec \phi$ in intensity formulas and $\lambda \sec^{1/2} \phi$ in duration formulas for high densities. The ½ exponent is used as a rough average between the values of 1 and 0.15 for along-the-path and across-the-path high-

TABLE 1. Summary of the theoretical characteristics of meteor scatter signals.

Characteristics	Long wavelength	Short wavelength
	Low density	
Echo intensity	$P_TG^2q^2R^{-3}\lambda^3\sec\phi$	$P_TG^2q^2D^{-2}v^2R^{-4}\lambda^6\sec^4\phi$
Echo duration	$D^{-1}(\lambda\sec\phi)^2$	$Dv^{-2}R(\lambda\sec\phi)^{-1}$
Echo energy	$P_TG^2q^2D^{-1}R^{-3}\lambda^5\sec^3\phi$	
Number with intensity $> P$	$P^{-0.5}$	
Number with duration $> \tau$	$\exp(-\tau)$	$(\tau^2+1)^{-1/2}$
Number of echoes above power level L	$P_T^{1/2}L^{-1/2}R^{1/2}(\lambda\sec\phi)^{3/2}$ [a]	$P_T^{1/2}L^{-1/2}D^{-1}v(\lambda\sec\phi)^3$ [a]
Average signal power	$P_TGD^{-1}R^{-1}\lambda^5\sec^4\phi$ [b]	
Median power level	$P_TD^{-2}R(\lambda\sec\phi)^7$ [c]	$P_Tv^{-2}R^2(\lambda\sec\phi)^4$ [c]
Duty cycle at power level L	$P_T^{1/2}L^{-1/2}D^{-1}R^{1/2}(\lambda\sec\phi)^{7/2}$ [d]	$P_T^{1/2}L^{1/2}v^{-1}R(\lambda\sec\phi)^2$ [d]
	High density	
Echo intensity	$P_TG^2q^{1/2}R^{-3}\lambda^3\sec\phi$	$P_TG^2q^{5/2}D^{-2}v^2R^{-4}\lambda^6\sec^2\phi$
Echo duration	$qD^{-1}\lambda^2\sec\phi$	$q^{-1}Dv^{-2}R\lambda^{-1}$
Echo energy	$P_TG^2q^{3/2}D^{-1}R^{-3}\lambda^5\sec^2\phi$	
Number with intensity $> P$	$P^{-2.0}$	$P^{-0.4}$
Number with duration $> \tau$	τ^{-1}	
Number of echoes above power level L	$G^{-1}R^2\sec\phi$ [e]	
Average signal power	$P_TGD^{-1}R^{-1}\lambda^5\sec^3\phi$ [e]	
Median power level	(small but complicated effect)	
Duty cycle at power level L	$G^{-1}D^{-1}R^2(\lambda\sec\phi)^2$ [e]	$G^{-1}Dv^{-2}R^3\lambda^{-1}\sec\phi$ [e]

[a] Plus number of high density.
[b] Plus average power of high density.
[c] Plus small effect of high density.
[d] Plus duty cycle of high density.
[e] Total of high density.

density trails. It is also assumed that the propagation path is symmetrical ($R_1 = R_2 = R$) and that the antennas at the two ends of the path illuminate the same small area of the meteor region to one side of the path mid-point ($G = 4\pi A\lambda^{-2}$). This illuminated area is proportional to $G^{-1}R^2\sec\phi$. An average value equal to $\sec\phi$ is used for the $(1 - \sin^2\phi\cos^2\beta)^{-1}$ variation in echo characteristics, so that the R in the radar formulas remains R for the symmetrical oblique paths. To show the effects of the various parameters, I is replaced by $G^2R^{-4}\lambda^2$. (Note that the λ's used here and in the antenna gain-aperture relation are not scaled by a $\sec^m\phi$ factor for oblique propagation.) The polarization angle α is assumed to be $90°$.

The first five rows in Table 1 need no comment since they repeat the information given in Sec. 3, with the foregoing substitutions.

In the row for the number of echoes above power level L, it is assumed that this threshold level is below the lowest peak intensity of echoes from high-density trails. Thus the number of echoes in the high-density columns is proportional to the illuminated area. The number of echoes from low-density trails is found by setting the expression for peak echo intensity equal to L, solving for the corresponding value of q, and using this value to find the integrated number (which is directly proportional to the illuminated area and inversely proportional to the threshold line density). Note in particular that the number of echoes from low-density trails is independent of antenna gain. That is, as antenna gain is increased, the effect on the number of echoes of increasing the sensitivity exactly compensates for the effect of decreasing the area of coverage. To find the total number of echoes above threshold L, the numbers from low- and high-density trails should be added. This cannot be done directly from the expressions given in the table, however, since these expressions show only functional behavior. Complete information on the system parameters is required for a calculation of the relative numbers of echoes from low- and high-density trails. For sensitive systems, most of the echoes will be due to low-density trails.

The average signal power is found in each case from the product of echo energy and area of antenna coverage. If the integrated number were inversely proportional to q over an infinite range of line densities, the average signal power would be infinite. The precise relative importance of the low- and high-density trails in the average signal power thus depends upon the limits of the meteor number-mass spectrum. It is believed that the average signal power is due primarily to the low-density trails.

The median power level is that power level above and below which the signal exists for equal periods of time. If only the low-density trails are considered, the median power level can be found by setting the product of number of echoes (area over line density) and echo duration equal to the constant value required to give a 50-percent duty cycle,[9] solving this expression for q, and using this value of q in the expression for echo intensity. The results are shown in Table 1. Note in particular the $R\lambda^7\sec^7\phi$ variation for the long-wavelength, low-density case. This is approximately the wavelength and distance dependence of the median signal level measured in ionospheric scatter propagation.[12, 13] It is believed that the effect of the high-density trails on the functional behavior of the median power level is slight, although their effect could be included from computations based on the parameters of a given system.

The duty cycle is of prime importance in a consideration of meteor-burst communication. The duty cycle is the fractional time the signal is above the power threshold L. It is assumed as before that L is well below the level of all echoes from high-density trails, so that the duty cycle due to these trails is proportional to the area of antenna coverage times the echo duration. For the low-density trails the product of number of echoes and echo duration is set equal to the duty cycle. Solving the resulting expression for q, substituting this into the echo intensity expression which is set equal to L, and solving for the duty cycle, yields the expressions in the table. (If the number and durations of echoes are so high that there is appreciable overlap, the true duty cycle is $1 - e^{-c}$, where c is the duty cycle as used above.[9]) The total duty cycle results from both low- and high-density trails. Their separate contributions could be computed for any given set of system parameters.

Note from Table 1 that duty cycle, median power level, and number of echoes above a threshold level are independent of antenna gain. Two simplifying assumptions were made in deriving this result. It is assumed that the antenna beams at transmitter and receiver illuminate each meteor trail with equal intensity. Also it is tacitly assumed that the common area in the antenna beams is uniform with respect to the number of trails per unit area oriented to provide propagation. But certain areas in the meteor region are much better for propagation than others, and the favored area changes position and shape diurnally and seasonally.[53] Thus the signal characteristics can be changed considerably by using high-gain antennas directed so as to maximize or minimize the meteor scatter signal.

It should be emphasized that the results given in Table 1 were obtained from a first-order analysis of the various factors which control meteor signal characteristics. Some of the indicated variations check better with experiment than others. It is sometimes difficult to separate combined effects in the measurements for close comparison with theory. For example, it is predicted in Table 1 that the duty cycle for low densities and long wavelengths is proportional to $L^{-1/2}$. However, measurements of the duty cycle at various levels for several oblique paths show exponents of L varying from about -0.6 to -1.3.[54, 58] It appears that the meteor trails in these investigations were primarily in the transitional region between low and high densities. Measurements of the number-intensity distribution lie in the transition region between the variations given by $P^{-0.5}$ and

TABLE 2. Summary of some characteristics of meteor scatter signals measured at Stanford University and the Stanford Research Institute.

f (Mc/s)	λ (m)	P_T (kw)	G_T (db)	G_R (db)	Path length (km)	Bandwidth (c/s) Predetection	Bandwidth (c/s) Postdetection	Signal Description
23	13	60	26	26	0	10,000	A-scope	b
23	13	60	9	9	0	10,000	A-scope	c
23	13	0.75	5	5	960	1000	100	d
46	6.5	0.60	10	10	960	1000	100	e
92	3.26	0.30	11	11	960	1000	100	f
40.4	7.45	2	9	9	1320	6000	100	g
61.3	4.9	100	ERP[a]	16	1030	6000	100	h
106	2.82	50	24	24	0	6000	A-scope	i
106	2.82	10	24	22	1320	6000	10	j
398	0.75	60	36	36	0	6000	A-scope	k

[a] 360° coverage in azimuth.

[b] Echo rates varying from as high as 10,000/hr in the early morning to a few hundred per hour in late afternoon. Threshold level is due to cosmic noise. The high-gain antenna used for transmission and reception is pointed north.

[c] Same echo rates as indicated in Ref. b above. In addition, echo rate is a function of the bearing of the antenna beam, as discussed in text.

[d] Average duty cycle 19.6 percent at about 3 db above cosmic noise level during early morning hours in September.[60]

[e] Average duty cycle 12.7 percent under conditions indicated in Ref. d.

[f] Average duty cycle 1.1 percent under conditions indicated in Ref. d.

[g] Meteor burst duty cycle varying diurnally between about 20 and 1 percent in August. Threshold level approximately 3 db above cosmic noise.[54]

[h] Meteor burst rate from television station KTVK in Phoenix, Arizona, varying diurnally between about 1200 and 50/hr in November–December. Optimum antenna bearing changes with time of day as predicted in text. Threshold level approximately 3 db above cosmic noise.[54]

[i] Echo rate at most favorable time of day during March through June averages 200/hr. Minimum detectable signal is about 10^{-16} w. Antenna azimuth and elevation angles adjusted for optimum meteor detection.[61]

[j] Echo rates varying diurnally from more than 1000/hr to about 100/hr at the optimum off-path antenna bearing. The optimum bearing changes up to $\pm15°$ from the direct bearing during the day. Minimum detectable signal is approximately 5×10^{-18} w.

[k] Echo rate at most favorable time of day during March through June averages 0.5/hr. Minimum detectable signal is about 10^{-16} w. Antenna azimuth and elevation angle adjusted for optimum meteor detection.[61]

$P^{-2.0}$. [25, 49, 58, 59] When the measurements are extended to smaller meteors by the use of more sensitive systems, it is believed that both the number of echoes and the duty cycle will vary approximately with the threshold level raised to the $-\frac{1}{2}$ power.

The theoretical functional behavior of the various signal characteristics are shown in Table 1. In Table 2 are described actual meteor signal characteristics measured at various frequencies and path lengths at Stanford University and the Stanford Research Institute. These descriptions are necessarily approximate, but they should serve as a rough indication of the actual signal characteristics for extrapolation by the functional relations to other sets of system parameters. (Note that when the threshold level is based upon cosmic noise, L is proportional to receiver bandwidth and to $\lambda^{2.3}$, approximately.)

6. Conclusions

Meteor scatter is an important mode of propagation at frequencies from about 20 to several hundred megacycles per second. From a consideration of the mechanism of radio-wave scattering by individual trails and the number and directional characteristics of meteor particles, it is possible to determine the characteristics of the total meteor scatter signal. From this knowledge, system parameters may be chosen to maximize the meteor signal (ionospheric scatter and meteor burst communication systems) or to minimize its effect (where the meteor scatter mode interferes with the desired operation of the system).

The author's research on meteor scatter is being sponsored by the Electronics Research Directorate of the Air Force Cambridge Research Center, Air Research and Development Command, under Contract AF19(604)–2193.

REFERENCES

1. A. C. B. Lovell, *Meteor astronomy* (Oxford University Press, New York, 1954).
2. P. B. Gallagher, "An antenna array for studies in meteor and radio astronomy at 13 meters," *Proc. I. R. E. 46,* 89 (1958).
3. L. A. Manning, O. G. Villard, Jr., and A. M. Peterson, "Meteoric echo study of upper atmosphere winds," *Proc. I. R. E. 38,* 877 (1950).
4. J. S. Greenhow, "A radio echo method for the investigation of atmospheric winds at altitudes of 80 to 100 km," *J. Atm. Terrestrial Phys. 2,* 282 (1952).
5. W. G. Elford and D. S. Robertson, "Measurements of winds in the upper atmosphere by means of drifting meteor trails II," *J. Atm. Terrestrial Phys. 4,* 271 (1953).

Willimantic State College Library

WILLIMANTIC, CONN

6. S. Evans, "Atmospheric pressures and scale heights from radio echo observations of meteors," *Meteors*, spec. suppl. to *J. Atm. Terrestrial Phys. 2,* 86 (1955).

7. D. K. Bailey, R. Bateman, L. V. Berkner, H. G. Booker, G. F. Montgomery, E. M. Purcell, W. W. Salisbury, and J. B. Wiesner, "A new kind of radio propagation at very high frequencies observable over long distances," *Phys. Rev. 86,* 141 (1952).

8. O. G. Villard, Jr., A. M. Peterson, L. A. Manning, and V. R. Eshleman, "Extended-range radio transmission by oblique reflection from meteoric ionization," *J. Geophys. Research 58,* 83 (1953).

9. V. R. Eshleman and L. A. Manning, "Radio communication by scattering from meteoric ionization," *Proc. I. R. E. 42,* 530 (1954).

10. D. W. R. McKinley, "Dependence of integrated duration of meteor echoes on wavelength and sensitivity," *Can. J. Phys. 32,* 450 (1954).

11. W. J. Bray, J. A. Saxton, R. W. White, and G. W. Luscombe, "VHF propagation by ionospheric scattering and its application to long-distance communication," *Proc. I. E. E. 103 B,* 236 (1956).

12. D. K. Bailey, R. Bateman, and R. C. Kirby, "Radio transmission at VHF by scattering and other processes in the lower ionosphere," *Proc. I. R. E. 43,* 1181 (1955).

13. O. G. Villard, Jr., V. R. Eshleman, L. A. Manning, and A. M. Peterson, "The role of meteors in extended-range VHF propagation," *Proc. I. R. E. 43,* 1473 (1955).

14. P. A. Forsyth and E. L. Vogan, "Forward-scattering of radio waves by meteor trails," *Can. J. Phys. 33,* 176 (1955).

15. Meteor Papers in *Proc. I. R. E. 45* (December 1957).

16. L. A. Manning, "The initial radius of meteoric ionization trails," *J. Geophys. Research 63,* 181 (1958).

17. F. L. Whipple, "Meteors and the earth's upper atmosphere," *Rev. Mod. Phys. 15,* 246 (1943).

18. L. A. Manning, O. G. Villard, Jr., and A. M. Peterson, "The length of ionized meteor trails," *Trans. Amer. Geophys. Union 43,* 16 (1953).

19. V. R. Eshleman, "The theoretical length distribution of ionized meteor trails," *J. Atm. Terrestrial Phys. 10,* 57 (1957).

20. J. S. Greenhow and E. L. Neufeld, "The diffusion of ionized meteor trails in the upper atmosphere," *J. Atm. Terrestrial Phys. 6,* 133 (1955).

21. G. H. Keitel, "Certain mode solutions of forward scattering by meteor trails," *Proc. I. R. E. 43,* 1481 (1955).

22. M. Loewenthal, "On meteor echoes from underdense trails at very high frequencies," Report No. 132, M.I.T. Lincoln Laboratory, December 1956.

23. A. C. B. Lovell and J. A. Clegg, "Characteristics of radio echoes from meteor trails I," *Proc. Phys. Soc. B 60,* 491 (1948).

24. N. Herlofson, "Plasma resonance in ionospheric irregularities," *Arkiv Fysik 3,* 247 (1951).

25. V. R. Eshleman, "The mechanism of radio reflections from meteoric ionization," Report No. 49, Stanford University, Contract N6onr–251, July 1952.

26. V. R. Eshleman, P. B. Gallagher, and R. F. Mlodnosky, "Meteor rate and radiant studies," Final Report, Stanford University, Contract AF16(604)–1031, February 1957.

27. T. R. Kaiser and R. L. Closs, "Theory of radio reflections from meteor trails, I," *Phil. Mag. 43*, 1 (1952).

28. M. E. Van Valkenburg, "The two-helix method for polarization measurements of meteoric radio echoes," *J. Geophys. Research 59*, 359 (1954).

29. L. A. Manning, O. G. Villard, Jr., and A. M. Peterson, "Radio doppler investigation of meteoric heights and velocities," *J. Appl. Phys. 20*, 475 (1949).

30. J. G. Davies and C. D. Ellyett, "The diffraction of radio waves from meteor trails and the measurement of meteor velocities," *Phil. Mag. 40*, 614 (1949).

31. D. W. R. McKinley, "Meteor velocities determined by radio observations," *Astron. J. 113*, 225 (1951).

32. V. R. Eshleman, "The effect of radar wavelength on meteor echo rate," *Trans. I. R. E.* PGAP-1, October 1953.

33. V. R. Eshleman, "Short-wavelength radio reflections from meteoric ionization. Part I: Theory for low-density trails," Report No. 5, Stanford University, Contract AF19(604)–1031, August 1956.

34. W. A. Flood, "Meteor echoes at ultra-high frequencies," *J. Geophys. Research 62*, 79 (1957).

35. E. R. Billam and I. C. Browne, "Characteristics of radio echoes from meteor trails. IV: Polarization effects," *Proc. Phys. Soc. B 69*, 98 (1956).

36. G. S. Hawkins, "Radar echoes from meteor trails under conditions of severe diffusion," *Proc. I. R. E. 44*, 1192 (1956).

37. J. S. Greenhow, "Characteristics of radio echoes from meteor trails. III: The behavior of the electron trails after formation," *Proc. Phys. Soc. B 65*, 169 (1952).

38. C. O. Hines and P. A. Forsyth, "The forward-scattering of radio waves from over-dense meteor trails," *Can. J. Phys. 35*, 1033 (1957).

39. L. A. Manning, "Oblique echoes from over-dense meteor trails," *J. Atm. Terrestrial Phys. 14*, 82 (1959).

40. L. A. Manning, "Air motions at meteoric heights," *Proceedings of the Mixed Commission of the Ionosphere*, August 1957.

41. J. S. Greenhow, "The fluctuation and fading of radio echoes from meteor trails," *Phil. Mag. 41*, 682 (1950).

42. L. A. Manning and V. R. Eshleman, "Discussion of the Booker and Cohen paper, 'A theory of long-duration meteor echoes based on atmospheric turbulence with experimental confirmation,'" *J. Geophys. Research 62*, 367 (1957).

43. H. G. Booker and R. Cohen, "A theory of long-duration meteor echoes based on atmospheric turbulence with experimental confirmation," *J. Geophys. Research 61*, 707 (1956).

44. D. W. R. McKinley and P. M. Millman, "A phenomenological theory of radar echoes from meteors," *Proc. I. R. E. 37*, 364 (1949).

45. G. S. Hawkins and D. F. Winter, "Radar echoes from over-dense meteor trails under conditions of severe diffusion," *Proc. I. R. E. 45*, 1290 (1957).

46. D. W. R. McKinley, "The meteoric head echo," *Meteors*, spec. suppl. to *J. Atm. Terrestrial Phys. 2*, 65 (1955).

47. F. G. Watson, *Between the Planets* (rev. ed.) (Harvard University Press, Cambridge, Massachusetts, 1956).

48. I. C. Browne, K. Bullough, S. Evans, and T. R. Kaiser, "Characteristics of radio echoes from meteor trails. II: The distribution of meteor magnitudes and masses," *Proc. Phys. Soc. B 69*, 83 (1956).

49. P. A. Forsyth, E. L. Vogan, D. R. Hansen, and C. O. Hines, "The principles of Janet—A meteor-burst communication system," *Proc. I. R. E. 45*, 1642 (1957).

50. G. S. Hawkins, "A radio echo survey of sporadic meteor radiants," *Monthly Notices Roy. Astron. Soc. 116*, 92 (1956).

51. C. O. Hines and R. E. Pugh, "The spatial distribution of signal sources in meteoric forward-scattering," *Can. J. Phys. 34*, 1005 (1956).

52. M. L. Meeks and J. C. James, "On the influence of meteor-radiant distributions in meteor-scatter communication," *Proc. I. R. E. 45*, 1724 (1957).

53. V. R. Eshleman and R. F. Mlodnosky, "Directional characteristics of meteor propagation derived from radar measurements," *Proc. I. R. E. 45*, 1715 (1957).

54. W. R. Vincent, R. T. Wolfrom, B. M. Sifford, W. E. Jaye, and A. M. Peterson, "Analysis of oblique path meteor-propagation data from the communications viewpoint," *Proc. I. R. E. 45*, 1701 (1957).

55. P. B. Gallagher and V. R. Eshleman, " 'Sporadic shower' properties of very small meteors," *J. Geophys. Research* (in press).

56. V. C. Pineo, "Off-path propagation at VHF," *Proc. I. R. E. 46*, 922 (1958).

57. K. L. Bowles, "Ionospheric forward scatter," *Ann. IGY 3*, pt. 4, 346 (1957).

58. H. J. Wirth and T. J. Keary, "The duty cycle associated with forward-scattered echoes from meteor trails," *I. R. E. Convention Record 6*, pt. 1, 127 (1958).

59. J. M. Taff and J. Damelin, "A preliminary survey of meteoric reflection of VHF signals recorded at Allegan, Michigan, during August and September, 1951," FCC report T.R.D. 2.1.5.

60. O. G. Villard, Jr., A. M. Peterson, L. A. Manning, and V. R. Eshleman, "Some properties of oblique radio reflections from meteor ionization trails," *J. Geophys. Research 61*, 233 (1956).

61. R. L. Leadabrand, Lambert Dolphin, and A. M. Peterson, "Upper atmospheric clutter research," Final Report, Contract AF30(602)–1462, Stanford Research Institute, October 1957.

5

Electromagnetic Emission
from Meteors

GERALD S. HAWKINS

A Search for Magnetic Effects from Meteors

Introduction

A. G. Kalashnikov[1] has published an extensive account of observations made in the Moscow area and near Garm in the North Pamirs of magnetic pulses which were attributed to meteors in the upper atmosphere. The fluxmeter used consisted of a sensitive moving-coil galvanometer connected to a large induction coil, composed of 10 turns, buried a few feet beneath the earth's surface. The coils varied in diameter from 100 to 300 m. With the galvanometer's sensitivity of 300 max/mm a deflection of 1 mm of the galvanometer's spot corresponded to a field of variation of 1×10^{-7} oersted or 0.01γ. A photographic recording technique was sometimes used in which a film moved at a speed of 4 mm/s in a direction perpendicular to the movement of the galvanometer's spot. Under these conditions magnetic-field changes of the order of 4×10^{-8} oersted supposedly could be detected.

Kalashnikov presented statistical evidence which indicated a correlation between the appearance of a meteor and a corresponding deflection of the fluxmeter. The epochs of the maxima of the Leonid, Geminid, Quadrantid, and Perseid streams were observed with the magnetometer technique, and an increase in the rate of magnetic pulses was found to correspond to the expected maximum hourly rates of meteors. A direct correlation was attempted during the Perseid shower in August 1950; of 169 meteors observed visually, 49 coincided with the appearance of a magnetic pulse within the time limit of ± 1 sec. These results indicated that magnetic pulses were more prevalent during the times of the major meteor streams and, further, that in some cases a direct correspondence existed between the appearance of a meteor and the generation of a magnetic disturbance.

A positive result of this kind is of great interest in the development of a theory for the meteoric process in the upper atmosphere, because it is extremely difficult to account for the existence of such magnetic effects on the basis of our present knowledge. Further observational data concerning this phenomenon are therefore important. Section A of this paper describes an experimental attempt to verify the results of Kalashnikov.

Equipment and Method of Experimentation

Our project set up magnetometers at Sacramento Peak, New Mexico, in the vicinity of the Super-Schmidt meteor cameras of the Harvard expedition, to provide correlations between the magnetic records and the precise photographic data. The operation continued over the period April 1956 to August 1957, except for a break during the winter months while the magnetic coils underwent considerable modifications.

Our equipment included an ultra-low-frequency amplifier with the passband (see Fig. 1) between 0.5 c/s and 10 c/s. The amplifier has been described previously by Aarons and Henissart[2] and Aarons,[3] and was similar to a conventional audio amplifier with high-value components in the resistance-capacitor coupling between stages. To achieve stability and a constant gain at these low frequencies it was found necessary to use storage cells for all power supplies. With careful

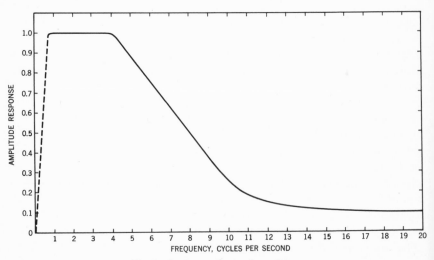

Fig. 1. Response curve of amplifier.

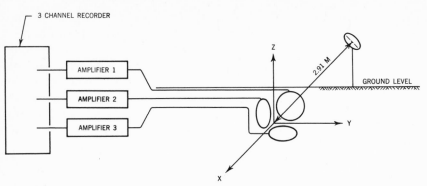

Fig. 2. Magnetometer coils and channels.

selection of the input tube the noise level of the amplifier could be maintained at a value equivalent to $1\mu v$ at the input. Three amplifiers of this type were lent the project by the Air Force Cambridge Research Center (AFCRC).

The first series of measurements used three coils to observe three components of the magnetic field. Each coil was 1m in diameter and contained 10,000 turns wound on a plastic former. An amplifier was connected to each coil and the outputs were fed to a three-channel recorder; each pen had a minimum response time of the order of 0.01 sec. To insure the minimum disturbance from earth tremors the three coils were buried in the rock, and to minimize interference, the hut containing the recording equipment was placed at a distance of 100 ft from the coils. The coils and three-channel recorder were also on loan from AFCRC. Dr. Curtis L. Hemenway and Dr. Jules Aarons supervised the initial construction and operation of this equipment.

A calibration coil 17 cm in diameter with 100 turns was placed at a distance of 2.91 m from the center of the three search coils, as shown in Fig. 2. The axis of the calibration coil made equal angles with the axis of each of the three search coils so that a current of 1 ma produced an equivalent normal field strength of 10^{-7} oersted through each of the search coils. Figure 3 shows a typical calibration record where the current in the calibration coil was altered abruptly following the form of a step-function. A response on the record equal in amplitude to the noise level would be produced by an abrupt change in magnetic field of $5 \times 10^{-2}\gamma$. The recorded noise level produced by the amplifier represented the limit of detection of the equipment. On some days, when the magnetic field of the earth was disturbed, the signal from the coils was greater than amplifier noise. The analysis did not include records obtained on days of magnetic disturbance.

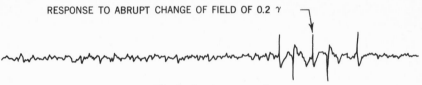

RESPONSE TO ABRUPT CHANGE OF FIELD OF 0.2 γ

Fig. 3. Specimen record.

To increase the sensitivity a new coil 100 m in diameter and containing 10 turns was laid 6 in. below the surface of the ground. A calibration coil consisting of one turn, 100 m in diameter, was laid in the same trench with the detection coil. The mutual inductance M between the two coils is given by Harnwell[4] as

$$M = \mu_o amn[\log_e(8a/d) - 2], \tag{1}$$

where a is the radius of the coils, d is their distance of closest approach, m and n are the number of turns, and μ_o is the permittivity of free space. Figure 4 shows a typical calibration record in which a current of 0.05 ma was reversed in the calibration coil every half-second. A Fourier analysis showed that the response of the amplifier to a 1-c/s field change of peak-to-peak amplitude $2A$ was equal to the response to an abrupt change of amplitude A. Thus an abrupt change of about $2 \times 10^{-3}\gamma$ would be at the limit of detection. The noise level shown in Fig. 4 is typical of that produced by variations of the earth's magnetic field under quiet conditions.

Many technical difficulties occurred in operating the equipment at this increased sensitivity. A visual observer, situated approximately 1 mi from the magnetic equipment, recorded the time of appearance of the meteor. The observer closed a circuit which actuated a side marker pen on the recorder. Although the current in the cable from the observer was reduced to 1 ma, this pulse was picked up by the detector coil. To overcome this and other problems, the hut containing the magnetic recording equipment was moved a distance of 200 ft from the edge of the detector coil and in this position the equipment worked satisfactorily.

RESPONSE TO FIELD VARIATION OF $3.4 \times 10^{-3}\gamma$ PER SECOND

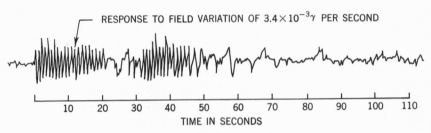

TIME IN SECONDS

Fig. 4. Specimen record.

Results

A detailed analysis was made of the characteristics of the meteor and any associated magnetic pulse. Table 1 summarizes the analysis for the first period of observation in which the low-sensitivity equipment was used. Results for each night of observation are given, and the meteors are divided into different brightness groups, on the basis of the observed visual magnitude. For each meteor an entry appears to show the amplitude of the associated magnetic noise pulse; the three digits in each group refer to the three coils. Any magnetic pulse that occurred within a period of 1 sec preceding or following the meteor was accepted as a correlation. A zero indicates that no magnetic pulse was recorded; when a pulse equal in magnitude to the noise level was found superimposed on the noise level, then "1" has been recorded. A "2" indicates a signal with amplitude 2×noise level. The table includes a column showing the magnetic pulses that were observed simultaneously with the time-check signals and are uncorrelated with meteors. An additional comparison sample was obtained by creating a random time series, which would not be expected to correlate significantly with the magnetic record. At the foot of the table appears the percentage of meteors associated with magnetic pulses greater in amplitude than the noise level, the corresponding percentage for time checks, and the random instances.

Table 2 records the magnetic results for the second period of observation with the equipment of increased sensitivity. Columns are included to indicate the magnetic pulses which correlated with time checks and with a sample of random instances selected to occur at approximately the same rate as meteors. As in Table 1, the amplitude of each pulse is given in units of the noise level. The percentage of correlation in each of these categories is recorded at the foot of the columns.

Analysis

The experiment with the low-sensitivity equipment showed that about 8 percent of meteors correlated with noise pulses; on the same basis, 7 percent of random instances correlated. There is an indication that the percentage of correlations increases with the brighter meteors, but the statistical significance is small. We are therefore led to conclude that the percentage of meteors producing magnetic pulses is only that to be expected from chance, and the experiment has produced no direct evidence that meteors generate magnetic pulses.

The second series of measurements, with the more sensitive equipment, yields a similar result. The percentage of meteors producing

TABLE 1. Pulses detected with low-sensitivity equipment.

Date 1956	Magnitude of meteor								Time checks	Random times
	3.1 to 4.0	2.1 to 3.0	1.1 to 2.0	0.1 to 1.0	−0.9 to 0.0	−1.0 to −1.9	−2.0 to −2.9	−3.0 to −3.9		
Aug. 6/7	000	001 110 001 {+ 11 / 000}	{11 / 000}	{3 / 000}	{2 / 000}	010 000	011	000	011 {+ 7 / 000}	
Aug. 7/8	{3 / 000}	111 {+ 9 / 000}	211 111 {+ 10 / 000}	{7 / 000}	{2 / 000}				111 {+ 27 / 000}	100 110 100 200 002 {+ 18 / 000}
Aug. 8/9	{3 / 000}	112 010 {+ 10 / 000}	210 {+ 18 / 000}	100 {+ 6 / 000}	010 000				{21 / 000}	002 010 010 {+ 11 / 000}
Aug. 9/10	102 {+ 3 / 000}	111 {+ 10 / 000}	202 {+ 7 / 000}	101 202 102 {+ 12 / 000}	000	{2 / 000}			{9 / 000}	
Aug. 10/11	001 000 000	110 102 {+ 18 / 000}	200 {+ 15 / 000}	202 200 {+ 9 / 000}	{4 / 000}	000		200	{31 / 000}	100 001 001 002 {+ 27 / 000}
Aug. 11/12		002 002 020 102 {+ 18 / 000}	010 {+ 34 / 000}	102 100 {+ 9 / 000}	002 {+ 10 / 000}	{3 / 000}	000		{20 / 000}	102 102 {+ 10 / 000}
Aug. 12/13	{2 / 000}	{18 / 000}	100 {+ 13 / 000}	200 {+ 13 / 000}	200 {+ 6 / 000}	000			{26 / 000}	
Aug. 13/14	{2 / 000}	200 200 020 002 {+ 8 / 000}	001 100 {+ 13 / 000}	{3 / 000}	003	000			{2 / 000}	

Coil	Meteors with associated pulse (percent)									
A	6	8	5	13	3	0	0	50	1	9
B	0	7	3	0	3	10	50	0	1	4
C	11	8	3	7	7	0	50	0	1	9

TABLE 2. Pulses detected with equipment of increased sensitivity.

Date 1957	Magnitude of meteor							No record of magnitude	Time checks	Random times
	4.1 to 5.0	3.1 to 4.0	2.1 to 3.0	1.1 to 2.0	0.1 to 1.0	−0.9 to 0.0	−1.0 to −1.9			
March 4/5	0		2,2 0,1 2	5,0 0,0	0,1 1,0				1,2 5,0 0,10 0,0 0,0 0,1	2,0 0,0 0,0 1,0 0,2 0,1 0,0 0,0 0,0 0,0 2,1
March 6/7			0	0					1,0 0	
March 7/8		0	3,0 0,1 0,1 1	3,2 0	0,0 1.5,2 2				0,0 1,1 1	0,0 0,0 1,1
April 24/25			0							
June 16/17		1	0	0		1		1,0	0,0 1	0,3 2
June 17/18		3	0					2	1,0	
June 18/19		0	0					0,0	0,2.5 0,0	
June 20/21		0	0					0,1		
June 21/22		1,1	0,0 1	0	0			0,0 0,0 0,0 1	0,0	0,0 0,0 1,1 0,0 0,0 1
June 23/24		0	0	0,0 0	1,1 0,0			0,0 0,1 1,1	0,0	1,0 0,0 0,1 0,0 1,0
June 25/26		1						0		
June 26/27	1	0	2					0,0 0,0	1,1	2,0 0,0 0,1 1

TABLE 2 (Continued)

Date 1957	Magnitude of meteor							No record of magnitude	Time checks	Random times
	4.1 to 5.0	3.1 to 4.0	2.1 to 3.0	1.1 to 2.0	0.1 to 1.0	−0.9 to 0.0	−1.0 to −1.9			
June 27/28		0,0 1	0,0 0	0,0 0,3 0,3	0		0	1,0 3,0	0,0 1,1 1,0 0	
July 5/6		2	0,1	3						
July 26/27			0,1							
Aug. 5/6					0				0,0	
Aug. 7/8		0	0,2	2,0 0	0				0,0 0,0	0,0 1,1 0,0 0
Aug. 8/9			0,1	0	0				1	
Aug. 25/26		2,0	2,1 1	0						

Percentages of meteors with associated pulse.										
	47	44	32	37			0	31	37	32

pulses is that to be expected if there were no correlation between meteors and the magnetic record. In the second series of experiments, however, it will be noted that the correlation has increased to 35 percent. This increase is attributed to the larger number of noise pulses detected with the high-sensitivity equipment, though these pulses cannot be directly attributed to meteors. The cause must be some other magnetic phenomenon. The difference in nature of the noise recorded at the two sensitivities is apparent in Figs. 3 and 4. The first equipment with a sensitivity of $5 \times 10^{-2}\gamma$ was limited for most of the time by amplifier noise, whereas the equipment with the sensitivity of $3 \times 10^{-3}\gamma$ was limited by the actual variations in the magnetic field of the earth.

At first sight this result appears directly to contradict the work of Kalashnikov. An analysis of his results, however, shows that the fraction of meteors producing detectable magnetic pulses was approximately 30 percent. This is comparable to the percentage found from the random time series in our preceding experiments, and it is therefore not unreasonable to assume that the correlations described by Kalashnikov had no significance, and that he was not justified in claiming that meteors produced magnetic effects. On the other hand, an inspection of his published records shows that the magnetic pulses detected with the Russian equipment were superimposed on an almost steady background. It could therefore be argued that the correlation of 35 percent was significant because there were no extraneous magnetic pulses. However, the magnetic noise level in the vicinity of Moscow is unlikely to be lower than the noise level at Sacramento Peak in New Mexico. If Kalashnikov's trace were indeed free from extraneous pulses for a large percentage of time, one would be forced to conclude that the noise level of his equipment was not as low as $3 \times 10^{-3}\gamma$, as stated. The general correlation, between the increase in the rate of occurrence of magnetic pulses and the increase in the hourly rate of meteors at the epochs of the major showers, remains unexplained. The correlation with the Leonid meteor stream is, however, immediately open to suspicion, for Leonid meteors have not been observed in any great numbers since 1932.

A Search for Radio Emission from Meteors

Equipment and Method of Observation

To search for possible radio emission from meteors, observations were made with three receivers having varying parameters as described in Table 3. The 218-Mc/s receiver was located at the Agassiz Station of Harvard College Observatory, Harvard, Massachusetts, and was constructed by the Ewen-Knight Corporation. A rotating switch compared the received signal with the noise in a resistor kept at constant temperature, and the receiver noise factor was measured

TABLE 3. Parameters of the equipment.

	Equipment 1	Equipment 2	Equipment 3
Frequency (Mc/s)	218	475	30
Bandwidth (Mc/s)	1	2	2
Integration time (sec)	1	1	1
Noise factor (db)	4	10	9
Minimum detectable flux (jan)	1.2×10^{-23}	1×10^{-22}	2×10^{-19}
Dish diameter (ft)	24	17	(half-wave dipole)
Beamwidth between 3 db points (deg)	12	8	40
Beamwidth between 1st minima (deg)	24	16	90
Observation period	Feb.–Apr. 1957	Dec. 1956–Aug. 1957	Dec. 1956–Aug. 1957

by using a noise diode incorporated in the front end of the receiver. A 24-ft paraboloid dish was used for an antenna, the center of the beam being directed in the region of the zenith. Radio signals were correlated with the observations of a visual observer. The over-all sensitivity of the equipment was determined by observing the radio source Cassiopeia A, on the assumption that the radio intensity from this source at the surface of the earth is 1.2×10^{-22} m^{-2} sec^{-1}.[5] The limiting sensitivity has been given in jansky units in Table 3, where 1 jan is equal to 1 w m^{-2}(c/s)$^{-1}$.

The second receiver, operating at a frequency of 475 Mc/s, was located at Mayhill, New Mexico, one of the camera sites of the Harvard Meteor Expedition. A 17-ft paraboloid dish was used for the antenna system and was directed so as to intersect the field of the Super-Schmidt cameras so that radio observations could be correlated directly with photographic data. An Esterline-Angus pen recorder was used to display the radio signal and the time of the appearance of the meteor was marked directly on this record from a timing device operated by a visual observer. The limiting sensitivity of the equipment was measured by use of radiation from the quiet sun and from the local galaxy, as reference signals.

The third receiver was also located at Mayhill, New Mexico, and operated in conjunction with the Super-Schmidt cameras.

Results

Observations at 475 Mc/s are summarized in Table 4, which records all meteors that passed within 12° of the center of the antenna beam. For example, on the night of December 10–11, 1956, three meteors were observed in the main beam of the antenna with apparent visual magnitudes of -0.5, $+2$, and $+3$. The numbers in the table refer to the amplitude of the radio signal in terms of the noise level. The meteors of magnitude -0.5 and $+2$ produced no observable signal, whereas the meteor of magnitude $+3$ was correlated with a radio impulse of twice the noise level. The noise level corresponds to a flux of 1×10^{-22} jan at the antenna. At times, meteors were observed outside the limits of the antenna beam; these are included in Table 4 under the heading "Out of field," and radio signals that correlated with these meteors are given. A similar tabulation is made in Table 5 for the observations with the 30-Mc/s equipment.

The observations at 218 Mc/s produced no signals coincident with meteors that were greater than one or two times noise. Details of the periods of observation and the magnitudes of the meteors that were within the half-power points of the antenna pattern are given in Table 6. Meteors recorded as "on the edge" were between the half-power point and the first minimum of the antenna pattern.

TABLE 4. Observations at 475 Mc/s.

Date	Magnitude of meteor											Meteors out of field
	−1	−0.5	0	+0.5	+1	+1.5	+2	+2.5	+3	+3.5	+4	
1956												
Dec. 2–3							0		0			0
Dec. 10–11		0					0		2			0
												0
Dec. 11–12				0			0					
							0					
							0					
							0					
Dec. 12–13							0		0			0
							0					
Dec. 13–14							0					0
							0					0
												0
												0
Dec. 27–28				0							1	0,2,
												0,2,
1957												
Jan. 4–5					0							
Jan. 20–21				0								
Jan. 24–25							0					0
												0
Feb. 26–27								0				
Mar. 4–5						0						0
												0
Mar. 7–8		0					0		0			
							0					
Mar. 8–9								0		0		0
May 19–20					0							
May 20–21									0			
May 23–24				0			0					
Mar. 5–6												0
Mar. 6–7												0
Aug. 7		0										0
												0
												0
Aug. 8–9												
Aug. 9–10												1
												0
Aug. 10–11							0					0
												0
												0
												0
												0
												0
Aug. 12–13	2						0					1
												1
												1
												1
												0
												0
												0
												0
												0
												0
												0
Total no. meteors	1	1	2	0	4	2	15	2	7	0	2	43
Total no. pulses	1	0	0	0	0	0	0	0	1	0	1	7

TABLE 5. Observations at 30 Mc/s.

Date	Magnitude of meteors								
1956	0	+0.5	+1	+1.5	+2	+2.5	+3	+3.5	+4
Dec. 12–13	1	15	1		1		0		
					1		1		
					0		1		
1957					1		0		
April 4–5				0	4	2	2	0	2
					0	0	0		
					0	0	0		
					0		0		
							0		
							0		
							0		
May 23–24	1		1		1	1	1		
			1				0		
			1				1		
Total meteors	2	1	4	1	9	4	14	1	1
Total pulses	2	1	4	0	5	2	5	0	1

Analysis

At a frequency of 218 Mc/s, 15 meteors passed through the center and 11 meteors through the edge of the antenna beam without producing any significant radio signal. It may be noted that the brightest meteor that passed through the center of the field was of magnitude +2. At a frequency of 475 Mc/s, a total of 36 meteors was seen to pass through the center or edge of the antenna field, and a few produced detectable pulses. In no case, however, was the signal greater than three times noise level and it is probable that the radio signal was not produced directly by the meteor. This view is supported by

TABLE 6. Observations at 218 Mc/s.

Date	Observing period EST	Meteor magnitudes	
		In field	On the edge
1957			
Feb. 3	0030–0430	+2	+2, +3, +3, +4
Feb. 4/5	0030–0545	+3, +3, +4, +4, +5	—
Feb. 6/7	0000–0445	+2, +2.5, +3	+3, +3
Feb. 8/9	2300–0300	—	—
Feb. 12	0200–0345	—	+1
March 4/5	2315–0330	+2, +4	+1
Apr. 21/22	0028–0515	+2, +3, +3, +3	+1, +2, +3

the fact that no significant tendency exists for the brighter meteors to produce more pulses than the fainter meteors. Furthermore, 16 percent of the meteors that were outside the antenna field were correlated with small radio pulses and presumably these pulses were of nonmeteoric origin. Thus the fact that 8 percent of the meteors within the beam were correlated with pulses is not significant.

A few pulses were found to occur simultaneously with the appearance of the meteor at a frequency of 30 Mc/s; one of these signals was 15 times noise. It is considered that the measurements at 30 Mc/s are not extensive enough to establish the validity of the correlation. Measurements at 30 Mc/s at the time of sunspot maximum are difficult to make, owing to the possibility of interference by scattering of man-made signals from the ionosphere. There is also the possibility that man-made signals may be reflected from the ionized column of the meteor. It is interesting to note that McKinley and Millman[6] described measurements that indicate an emission level from meteors some 30 db greater than reported here.

The evidence therefore indicated that meteors do not emit appreciable radio noise over the 30–475-Mc/s frequency range. At 218 Mc/s any radio flux from a meteor of visual magnitude 0 must be less than 1×10^{-22} jan, and at a frequency of 218 Mc/s this flux is less than 1.2×10^{-23} jan. If we assume a range of 150 km for the meteor, then the upper limit to the total power emitted over a pass-band of 1 Mc/s is as follows: at 30 Mc/s, 5.6×10^{-2} w; at 218 Mc/s, 3.3×10^{-6} w; at 475 Mc/s, 2.8×10^{-5} w. For comparison it is interesting to note that a zero-magnitude meteor, during the ionization process in the atmosphere, generates approximately 1.6×18^8 w.

It may be concluded that meteors do not emit radio noise that can be detected with conventional radiometers over the 30–475-Mc/s frequency range and that the radio emission in a passband of 1 Mc/s is less than 10^{-10} of the original kinetic energy. On the basis of these measurements there seems to be no possibility of a plasma oscillation occurring in the ionized column of the meteor trail.

Dr. Curtis L. Hemenway assisted in the construction of some of the equipment at Mayhill and the observations at Mayhill were carried out with the assistance of Messrs. M. L. Shapiro, R. Aikens, and G. Schwartz. At Agassiz Station the observing team included Messrs. R. B. Southworth and M. L. Shapiro. The Ewen-Knight radiometer was kindly made available by Dr. Jules Aarons of the Air Force Cambridge Research Center.

This research was sponsored jointly by the United States Army, Navy, and Air Force under contract with the Massachusetts Institute of Technology.

Discussion

P. A. Goldberg, *Boeing:* I should like to add a note of confirmation to Dr. Hawkins' paper. In the summer of 1955, at the University of Oregon we carried out some magnetometer measurements in the range 0.3–150 c/s, with equipment similar to that used by Dr. Hawkins. Three visual observers sat up through many weary hours of the night trying to find Kalashnikov's pulses on our magnetogram records. We can say this: when there are lots of little meteors, you can certainly see a correlation, but when a big one comes across, it just does not show up on the magnetic records. We concluded, therefore, that Kalashnikov had some kind of statistical difficulty with his problem. Anyone who is familiar with the appearance of these magnetogram records will know that they are full of wiggles and bursts and other kinds of impulses and signals, and you cannot pick a meteor out of that.

G. S. Hawkins: My own conclusion is that the correlations found in the Kalashnikov experiments are not significant. Nor is there any indication that the correlations increase when one deals with brighter meteors. On the other hand, Kalashnikov mentions a magnetic pulse received from a meteor of visual magnitude −8, which is considerably brighter than the brightest meteor described here. It is still possible that observable pulses are produced by fireballs, bolides, and meteorites.

V. R. Eshleman, *Stanford University:* We at Stanford also looked about seven years ago with a large coil, and we found no correlation with the Perseids that year.

Hawkins (*added in proof*): Some enhancements of low-frequency magnetic noise were observed in 1959 at the time of the Geminid meteor stream.[7]

REFERENCES

1. A. G. Kalashnikov, Academy of Science, USSR, No. 6 (1952).
2. J. Aarons and M. Henissart, *Nature 172,* 682 (1953).
3. J. Aarons, *Proc. Amer. Acad. Arts Sci. 79,* 266 (1953).
4. G. P. Harnwell, *Principles of Electricity and Electromagnetism* (McGraw-Hill, New York, 1938).
5. R. Adgie and F. G. Smith, *Observatory 76,* 181 (1956).
6. D. W. R. McKinley and P. M. Millman, *Proc. I. R. E. 37,* 364 (1949).
7. A. W. Jenkins, C. A. Phillips, and E. Maple, *J. Geophys. Research 65* (1960). In press.

6

Whistler-Mode Propagation

R. A. HELLIWELL

Introduction

Radio noise below about 25 Mc/s arises mainly from terrestrial sources. Near inhabited areas man-made sources often predominate; they include power lines, automobile ignition systems, fluorescent lights, and other devices. Elsewhere the most important source of noise is natural lightning. Lightning flashes produce electromagnetic impulses with a broad spectrum which peaks in the vicinity of 10 kc/s. The peak power in a single flash is of the order of 10^9 w.

The energy radiated by lightning covers a very wide frequency range, from as low as a few cycles to tens of megacycles per second. Lightning energy, like ordinary radio signals, reaches a receiver by all of the well-known mechanisms of propagation, including surface wave, tropospheric wave, and ionospheric sky wave. The intensity of atmospheric noise varies widely with time, location, and frequency.

The conventional mechanisms of propagation mentioned above are not the only means by which lightning energy can reach the receiver. At very low frequencies, below approximately 35 kc/s, appreciable amounts of lightning energy can reach a receiver by means of the whistler mode of propagation.[1] In this mode energy is guided by the lines of force of the earth's magnetic field. When lightning energy reaches the lower edge of the ionosphere a portion is reflected to cause conventional static, and a portion is transmitted into the ionosphere. Part of this energy is absorbed and part goes on, following the lines of force of the earth's field. If the electron density of the outer ionosphere is sufficient, the energy will be guided back to the earth in the opposite hemisphere. The signal which emerges is usually a gliding tone lasting 1 to 2 sec, and hence is called a "whistler." The important point is that the whistler may equal or exceed the direct signal in amplitude. The whistler thus becomes an important source of noise in the very-low-frequency region.

It is the purpose of this chapter to describe the basic theory of whistler-mode propagation and to review the available experimental knowledge of its properties.

Theory of Whistler-Mode Propagation

The principal properties of whistler propagation can be derived from the Appleton-Hartree expression for refractive index in an ionized medium containing a static magnetic field. The formula for the complex refractive index is given in conventional notation by

$$n^2 = 1 - \cfrac{x}{1 - jz - \cfrac{\frac{1}{2}y_T^2}{1 - x - jz} \pm \left[y_L^2 + \left(\cfrac{\frac{1}{2}y_T^2}{1 - x - jz} \right)^2 \right]^{1/2}} \tag{1}$$

where $x = (\omega_o/\omega)^2$, $\omega_o^2 = Ne^2/mk_o$, $N\,(\mathrm{m}^{-3})$ is the number of electrons, e (coul) is the charge on the electron, m (kg) is the mass of the electron, $k_o = 8.85 \times 10^{-12}$ f/m, ω is the angular wave frequency, $z = \nu/\omega$, ν is the collision rate, $y_T = \omega_T/\omega$, $y_L = \omega_L/\omega$, $\omega_T = eB_T/m$, $\omega_L = eB_L/m$, B_T is the component of the earth's field transverse to the direction of propagation, and B_L is the component of the earth's field parallel to the direction of propagation.

Whistler propagation is approximately described by the longitudinal case in which the quanity y_T is 0. Equation (1) becomes, in the absence of losses ($z = 0$),

$$n^2 = 1 - \frac{x}{1 \pm y_L}. \tag{2}$$

For very-low-frequency waves to be propagated through the ionosphere the refractive index n must be a real number. For the relatively large values of x usually encountered, this will be true only if we choose the minus sign in expression (2). This assumes that the quantity y_L is always greater than 1, which is generally the case for very-low-frequency propagation through most of the ionosphere. This case is sometimes called the "longitudinal extraordinary mode" and is given by

$$n^2 = 1 + \frac{\omega_o^2}{\omega(\omega_H - \omega)}. \tag{3}$$

Equation (3) shows that for values of f_H greater than f, n^2 will be greater than 1. As an example, we may calculate the refractive index of a 10-kc/s wave passing through the E-region of the ionosphere where the plasma frequency is 3 Mc/s and the gyro-frequency is 1.5 Mc/s. From Eq. (3) we find the refractive index to be approximately 24.5. The lower ionosphere is thus a medium of high refractive index at very low frequencies. Applying Snell's law to the lower boundary, we find that all incident waves will excite transmitted waves whose wave normals are contained within a small angle about the surface normal. The result is a pronounced "beaming" of the energy in a direction at right angles to the planes of stratification. The anisotropy of the medium results in bending of the beam in the direction of the earth's field.

To understand the frequency-time behavior of whistlers it is necessary to calculate the group velocity, which is related to the refractive index by

$$v_g = c \left(n + f\frac{dn}{df} \right)^{-1}, \tag{4}$$

where c is the velocity of light.

From (3) and (4), plus some manipulation, we can obtain the expression for group velocity which has been reported elsewhere.[2] When $\omega_o{}^2$ is large compared with the product $\omega\omega_H$, as is usually the case, we can obtain a simplified expression for group velocity, namely

$$v_g = 2c\frac{f^{1/2}(f_H - f)^{3/2}}{f_H f_o}. \tag{5}$$

This is the equation that explains the "nose whistler," in which the originating impulse is transformed into simultaneous rising and falling tones. From Eq. (5) it is easily found that the group velocity is a maximum when $f = \frac{1}{4}f_H$. This is called the nose frequency. Expression (5) shows that, if f exceeds f_H, the group velocity becomes imaginary. Thus f_H becomes the maximum usable frequency for the whistler mode of propagation. In practice, this means that the upper cutoff frequency, or maximum usable frequency, will be the minimum value of the gyro-frequency encountered over the path of propagation. If, as is usually assumed, the path of propagation is along the lines of force of the earth's magnetic field, the upper cutoff frequency will be the value of the gyro-frequency at the top of the path. This quantity will be a function of the geomagnetic latitude of the path of propagation. The value of the minimum gyro-frequency drops off inversely with the cube of the distance from the center of the earth. At 20°, 40°, and 60° geomagnetic latitude, the corresponding cutoff frequencies are 630 kc/s, 165 kc/s, and 13.5 kc/s, respectively.

At medium and low latitudes and at frequencies below about 10 kc/s Eq. (5) can be usefully approximated by

$$v_g = 2c \frac{(f f_H)^{1/2}}{f_o}, \tag{6}$$

which is the form developed by Eckersley to explain the frequency-time behavior of whistlers.

For a path of constant length, Eq. (6) shows that a graph of $f^{-1/2}$ as a function of t should be a straight line. Most low-frequency whistlers fit this law very closely. By going backward in time one can see that a signal which rises in frequency according to Eq. (6) should at some point along the path create an impulse. Further along the path the signal will transform into a descending tone. As a whistler bounces back and forth between the hemispheres the dispersion will increase according to Eq. (5) or (6). For a given path, the time delays of successive whistler echoes should be in the ratios $1:3:5:7: \cdots$ when the source is in the hemisphere opposite to that of the observer, and $2:4:6:8: \cdots$ when source and observer are in the same hemisphere. The former whistlers are called "short" and the latter "long."

The guiding of whistlers results from the anisotropic behavior of the transmission medium. Detailed consideration of Eq. (1) shows that the refractive index is a strong function of the direction of propagation relative to the direction of the earth's field. The direction of energy propagation called the "ray path" is different from the direction of the wave normal and it is necessary to perform a rather involved analysis in order to obtain the actual ray path. It has been shown by Storey[3] that when Eq. (6) applies the ray path will be limited to directions within a cone of semiangle $19°26'$. The actual path of the ray will depend upon the gradient of refractive index.

The attenuation rate in the whistler mode is, of course, important in noise calculations. A first approximation to the attenuation rate can be obtained from Eq. (1) for the longitudinal case. For the case when losses are small, so that z^2 is small compared with the quantity $(1 - y_L)^2$, it can be shown that the attenuation rate is given by

$$\alpha = \frac{\nu \omega_o}{2c \omega_H^{3/2}} \omega^{1/2}. \tag{7}$$

The important point is that the attenuation rate is proportional to the square root of frequency. This law contrasts markedly with that for ordinary high-frequency ionospheric propagation in which the attenuation rate is approximately inversely proportional to the square of the frequency. It is thus clear that for whistlers there will be an upper cutoff frequency resulting simply from ordinary energy dissipation in the lower regions of the ionosphere.

Observations

The empirical picture of whistler phenomena is far from complete, and only some preliminary results can be presented at this time. It is known that whistlers can be observed from a minimum frequency of about 400 c/s to a maximum frequency of about 35 kc/s. It is not yet known to what extent the spectrum of whistlers is controlled by the spectrum of the lightning source and to what extent it is controlled by propagation factors. The peak energy in most whistlers seems to lie in the range 3 to 10 kc/s.

The time delays of whistlers vary over a large range. The dispersion constant D sec$^{1/2}$ will typically range from 20 to 200 ($D = tf^{1/2}$). An example of a middle-latitude whistler recorded at Stanford, California, and Boulder, Colorado, is shown in Fig. 1. This is a short whistler with multiple components and several echoes. At Stanford the whistler can be traced to 32 kc/s and at Boulder a distinct but weak nose component can be seen with a nose frequency of about 15 kc/s. The vertical lines on the figure represent the conventional sferics which reach the receiver by propagation beneath the ionosphere. This example illustrates in a qualitative way the relative importance of whistler energy compared with energy received directly by means of propagation beneath the ionosphere. The relative importance of

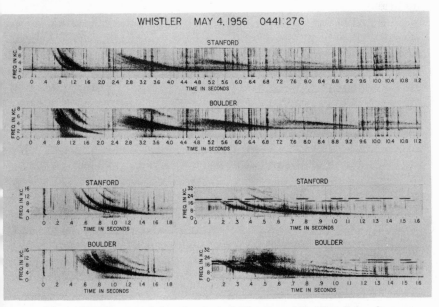

Fig. 1. Whistler recorded at Stanford, California, and Boulder, Colorado.

whistlers and sferics in a particular receiver is difficult to assess because of the limiting effects which often occur in the presence of
strong atmospherics. It can safely be said, however, that if an audio-
frequency receiver is adjusted to maximum sensitivity, there will frequently be times when the background noise is produced mainly by
whistlers.

Whistlers have not been heard near the geomagnetic equator nor
at the geomagnetic poles. They occur most frequently in middle latitudes, and they are more frequent during night than day. There has
been reported a positive correlation between whistler occurrence and
magnetic activity,[3] although this has not been substantiated by recent
measurements. Some idea of the relative importance of the whistler
mode as compared with direct propagation beneath the ionosphere
can be found in tests recently performed on Station NSS.[4] Pulses were
transmitted between Annapolis and Cape Horn, South America, on
15.5 kc/s. At night a whistler-mode echo was observed with amplitudes running between 10 and 30 db below the amplitude of the direct
wave. Because the receiver was about 1000 km away from the calcu-
lated conjugate position, it is probable that the observed whistler-
mode echo was much weaker than it would have been at the optimum
location. It seems likely that the strength of the whistler-mode signal
may in fact equal or exceed that of the direct signal under favorable
conditions.

The significance of whistlers in the determination of atmospheric
noise in a statistical sense is of course as yet impossible to measure. It
is possible, however, that noise maps constructed for frequencies of
15 kc/s or less will show a correlation between active noise centers in
one hemisphere and noise peaks in the region of the conjugate point
in the opposite hemisphere. As a matter of fact, it might be possible
to test the hypothesis that whistlers can contribute appreciable
amounts of noise by simply making a statistical comparison of primary
noise centers and the variation of noise in the vicinity of their conjugate points. This would have to be done at frequencies no greater
than 15 kc/s, with the optimum frequency probably lying in the
vicinity of 5 kc/s.

Recent research on very-low-frequency emissions[5] suggests the
possibility that during a magnetic disturbance whistlers may be am-
plified through the mechanism of a giant traveling-wave tube located
in the outer ionosphere. Streams of charged particles from the sun
perform the function of the electron beam in a tube. The ambient ionization in the presence of the earth's field is the slow-wave structure
of the medium. This would mean that energy can be added to the

noise generated in the troposphere. Thus it might be expected that the correlation between noise centers in one hemisphere and the noise near the conjugate points would be enhanced during periods of magnetic disturbance.

Discussion

E. W. Allen, *FCC:* Is the frequency dependence of whistlers such that they might affect the radio navigation frequency bands near 100 kc?

R. A. Helliwell: We do not know the efficiency of the mode at 100 kc/s but evidence is increasing that the cutoff is quite a bit lower. The causes of the cutoff are not known. A first approximation for the lower regions of the upper atmosphere indicates that the attenuation due to collisions increases in proportion to the square root of the frequency. This will tend to wipe out the higher frequencies.

D. H. Menzel, *Harvard College Observatory:* In the case of man-made whistlers, is the effect closely similar from day to day?

Helliwell: No, and this is one of the interesting things found in that experiment. The signal was received at Cape Horn almost all the time, whereas whistlers showed a very marked day-to-day variation in occurrence, which seems to indicate that the presence of whistlers is controlled mainly by thunderstorm activity in the appropriate geographic location.

P. A. Goldberg, *Boeing:* Can you say anything more about the effects of strong solar events on whistler propagation? This would be important in connection with the termination of the magneto-ionic duct, that is, the bottom of the D-region where the anisotropy disappears, but there may still be a significant amount of loss, especially during solar flares.

Helliwell: The phenomena during a disturbance are quite complex. It appears that there are periods when whistlers disappear. But there is no simple correlation of the occurrence of whistlers with magnetic activity. On the other hand, the presence of long trains of echoes seems to be greatest during periods of magnetic activity. So we can not answer this question completely yet, although it is a very important one.

REFERENCES

1. R. A. Helliwell and M. G. Morgan, "Atmospheric whistlers," *Proc. I. R. E. 47,* 200 (1959).
2. R. A. Helliwell, J. H. Crary, J. H. Pope, and R. L. Smith, "The 'nose' whistler, a new high latitude phenomenon," *J. Geophys. Research 61,* 139 (1956).

3. L. R. O. Storey, Ph. D. dissertation, Cambridge University, England, and *Phil. Trans. Roy. Soc. A 246,* 113 (1953).

4. R. A. Helliwell and E. Gehrels, "Observations of magneto-ionic duct propagation using man-made signals of very low frequency," *Proc. I. R. E. 46,* 785 (1958).

5. R. M. Gallet and R. A. Helliwell, "Origin of 'very-low-frequency emissions,'" *J. Research, Natl. Bur. Standards 63D,* 21 (1959).

7

The Radio Spectrum of
Solar Activity

A. MAXWELL, G. SWARUP,
and A. R. THOMPSON

Introduction

The radio emission from the sun is composed of a background thermal emission from the solar atmosphere, and bursts of radiation, sometimes very intense, which originate in localized active areas on the disk. These bursts have complex characteristics, which make their classification very difficult from observations made with single-frequency receivers. Observations over a large continuous band of frequencies, however, have made it possible to replace earlier empirical classifications of the radio bursts with a simple and natural classification based on their essential physical characteristics. The swept-frequency technique, in which a narrow-band tunable receiver is repeatedly swept across a wide frequency range, was originally used by Wild and McCready[2] for solar observations in the range 70–130 Mc/s, and subsequently by Wild, Murray, and Rowe[3] in the range 40–240 Mc/s.

This paper is concerned with a new set of observations which have been taken at the Harvard Radio Astronomy Station, Fort Davis, Texas, since September 1956. The new equipment covers the range 100–580 Mc/s. It uses an antenna with five times the collecting area of that used by Wild and his colleagues, and its receiver noise figures are approximately 4 db lower. Because of its large frequency range and high sensitivity, the equipment is exceedingly vulnerable to interference, especially in the lowest 200 Mc/s of its band, which is crowded with radio transmissions. It was therefore deemed necessary to set up the station in a remote part of the United States. A number of places were extensively surveyed for radio interference, and the site finally chosen was in the Davis Mountains of West Texas. The

Fig. 1. Laboratory and 28-ft paraboloid antenna near Fort Davis, Texas.

Station (Fig. 1) is situated in a broad valley, and is surrounded by mountains 1500 ft higher; these give horizons of about 5° elevation at all azimuths. At frequencies above 35 Mc/s, the station is virtually free from interference, apart from weak television signals on channels 2, 4, and 7, which have field strengths equivalent to 3, 1, and 4 μv at the terminals of a half-wave dipole; these are transmitted from the nearest large towns which are over 160 miles away.

Observations are taken daily from sunrise to sunset, and less than 5 percent of the possible observing time has been lost. The Station works in close cooperation with the Sacramento Peak Observatory, which forms its optical counterpart.

Equipment

The frequency range of the equipment, 100–580 Mc/s, is covered in three octave bands, 100–180, 170–330, and 300–580 Mc/s, by separate receivers. These sweep concurrently at a rate of 3 times/sec, and are connected to separate broad-band primary arrays at the focus of a paraboloid antenna. The outputs are displayed on three intensity-modulated cathode-ray tubes and recorded photographically.

The antenna has a diameter of 28 ft and a focal length of 12 ft, and is equatorially mounted. The primary arrays designed by Jasik are described elsewhere.[4] The low-frequency array is a double dipole and reflecting screen, the mid-frequency array a single dipole mounted in a corner reflector, and the high-frequency array an electromagnetic horn. They are mounted coaxially, and the mid-frequency array is cross-polarized with the other two. The effective collecting area of the antenna varies between 38 and 45 m^2 over the frequency range; this corresponds to an efficiency between 67 and 79 percent, the geometric aperture being 57 m^2.

The receivers were built by Airborne Instruments Laboratory to meet the specifications of the Station. They are described in detail in another paper.[5] In each receiver the r.f. head consists of two grounded-grid amplifier stages, an oscillator, and a crystal mixer. The local oscillator and the amplifier plate circuits are tuned by variable capacitors. These are geared to a synchronous motor and rotate at 3 rev/s. The full frequency sweep is achieved in 180° rotation, and the receiver output is rejected over the following half-cycle where it is not possible to maintain alignment. The noise figures are 6, 7, and 8 db for the low-, middle-, and high-frequency bands, respectively, and are achieved by the use of Western Electric 416B triodes in the r.f. stages. The i.f. bandwidths are 0.3, 0.5, and 1.0 Mc/s, that is, about ⅓₀₀ of the range covered in each case.

The display unit comprises three high-resolution cathode-ray tubes (DuMont K1207) which have a resolution of approximately 1000 lines across the 5-in. screen. A single vertical trace is produced on each tube by a deflection waveform which is derived from a discriminator fed by the local oscillator of the associated receiver; the spot displacement is therefore proportional to the instantaneous frequency of the receiver. The brightness is proportional to the receiver output. The three tubes are mounted one above the other, and are photographed by a continuous-motion 35-mm camera in which the film moves horizontally at a speed of the order of 12 mm/min. The pattern on the film is therefore an intensity-modulated graph of frequency against time. The film used, Kodak Linagraph Panchromatic, was chosen to give the best compromise between speed and grain size. The whole system responds to intensity variations over a range of 30 db.

Operation and Calibration of the Equipment

The equipment is switched on automatically at sunrise by a time clock. Minute marks are photographed on the films, and these are calibrated to an absolute accuracy of ± 1 sec from the WWV transmissions. The films are developed to photometric standards.

After sunset each day the equipment is calibrated in frequency and intensity. For the frequency calibration, signal generators are used to provide six chosen frequencies within each octave band. Thus by interpolation one may deduce the frequency of the solar radio signals at any point on the film.

For the intensity calibration, the output of a noise diode source is fed into each of the receivers at six different levels covering a range of 18 db. The film density is thereby calibrated in terms of noise power applied to the receiver. It then remains to relate this input power to the intensity of the solar radiation incident on the antenna. This may be done as follows.

Let N be the receiver input power corresponding to an antenna temperature T_A. The quantity N is a noise power, and is most conveniently expressed here in units of $kT_0\,\Delta f$ (the power available from a matched resistor at ambient temperature), where k is Boltzmann's constant, T_0 is the ambient temperature, and Δf is the noise bandwidth of the receiver. The power applied to the receiver input may also be expressed in terms of T_A and the attenuation α in the antenna cable.

Thus

$$NkT_0\,\Delta f = [\alpha T_A + (1 - \alpha)T_0]k\Delta f,$$

whence

$$T_A = \frac{1}{\alpha}(N + \alpha - 1)T_0. \tag{1}$$

The power delivered from the antenna terminals, $kT_A\,\Delta f$, is equal to that of the radiation absorbed by the antenna, that is,

$$kT_A\,\Delta f = \frac{1}{2}\,\Delta f \iint_{4\pi} AB d\Omega.$$

Here A is the effective collecting area, and B the brightness in w m^{-2} $(c/s)^{-1}$ sterad^{-1} which is absorbed by the antenna. Both are functions of the direction of the solid angle $d\Omega$, and the factor ½ is introduced because the antenna absorbs radiation in one plane of polarization only. The integral may be evaluated in two separate parts, the first over the radio disk of the sun and the second over the remaining sky covered by the antenna beam. If B_1 is the effective average value of the brightness over this remaining section of sky, then

$$kT_A = \frac{1}{2}A_0 \iint_{\text{sun}} Bd\Omega + \frac{1}{2}B_1 \iint_{4\pi-\text{sun}} Ad\Omega. \tag{2}$$

It has been assumed that the antenna beam is wide compared with the sun, so that for the first integral, A, the effective collecting area, may be taken as constant and equal to A_0, the value corresponding to the center of the beam. On the same assumption the second integral is approximately that obtained by evaluating it over 4π sterad. (This approximation is correct to ¼ percent at 100 Mc/s, and to 3 percent at 580 Mc/s for the present equipment.)

Now,

$$\iint_{\text{sun}} B\,d\Omega = I, \tag{3}$$

and

$$\iint_{4\pi} A\,d\Omega = \lambda^2 \tag{4}$$

where I [w m^{-2}(c/s)$^{-1}$] is the intensity of solar radiation and λ is the wavelength of the radiation. Also B_1 is related to the equivalent value of temperature T_1 by

$$B_1 = \frac{2kT_1}{\lambda^2}. \tag{5}$$

Combining Eqs. (1) through (5),

$$I = \frac{2k}{A_0}\left[\frac{1}{\alpha}(N + \alpha - 1)\,T_0 - T_1\right].$$

Thus I, the intensity of the solar radiation, is determined in terms of N, which is directly related to the film density by the calibration. Of the other quantities involved, α and A_0 are known parameters of the equipment, $T_0 = 300°$K, and T_1 is the sky background temperature. (It is generally more convenient to express the sky radiation in terms of temperature than of brightness. The values of T_1 depend on the sun's position in the sky; those values given in Table 1 indicate the order of magnitude.) From the value of I it remains only to subtract the component due to the quiet sun to determine the intensity of the solar activity.

Approximate minimum detectable levels of solar activity at four frequencies are given for the present equipment in Table 1.

Observations

The early spectral observations made by Wild and his co-workers established the fact that a great majority of the solar radio bursts belong to one of four distinct types. The classification is natural in the sense that the various types can be interpreted in terms of different physical processes on the sun. The present observations confirm that

TABLE 1. Levels of solar radiation at four frequencies

Frequency (Mc/s)	Average sky temperature T_1 (°K)	Intensity of quiet sun at solar maximum $[10^{-22} \text{ w m}^{-2}(\text{c/s})^{-1}]$	Minimum detectable activity above quiet sun $[10^{-22} \text{ w m}^{-2}(\text{c/s})^{-1}]$
125	2000	4	5
200	500	10	10
425	100	30	20
550	50	40	20

these identifications account for more than 95 percent of the observed phenomena, but the greater sensitivity and different frequency range of the present equipment have given further information about the spectrum of the bursts. In this section the characteristics of the spectral classes are summarized briefly, and illustrated with examples from the records.

For the frequency band 100–580 Mc/s, the assumption may be made that the solar radiation at a given frequency originates near a level in the solar atmosphere where the plasma frequency corresponds to that of the radiation. One may thus transform the frequency band 100–580 Mc/s into a height range, which, at the present maximum of the solar cycle, lies roughly between 500,000 and 50,000 km above the solar photosphere. It is therefore possible to interpret the frequency coordinates on the film records in terms of height. However, it must be emphasized that, although this interpretation gives the best physical picture so far suggested, and although it holds for the quiet (thermal) solar radiation, its validity is by no means established for the case of violent outbursts.

Noise Storms

A noise storm consists of a long series of short bursts continuing over hours or days. They are superimposed on a background of slowly varying enhanced radiation which has been described as a "continuum," although it is possible that the background may itself be composed of a large number of overlapping bursts. Noise storms are normally spread over a large frequency band but are rarely seen above 250 Mc/s. On many occasions the bursts have bandwidths of a few megacycles and lifetimes extending from a fraction of a second to nearly 1 min (Fig. 2). At other times bursts of bandwidth nearly 30 Mc/s and lifetimes less than a second may predominate.

Noise storms are caused by localized disturbances in the solar corona above active sunspot regions, but details of the radiation processes are not clearly understood.

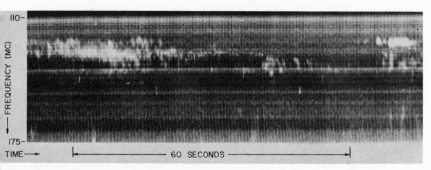

Fig. 2. Noise storm: 1958 October 30, 1355–56 U. T. The enlargement of this record is about 3 times greater than that of Figs. 3 and 4. The individual cathode-ray tube traces give the fine vertical structure.

Slow-Drift Bursts

A slow burst appears as a narrow band of intense radiation which drifts gradually, and sometimes irregularly, toward lower frequencies. Figure 3 shows a typical example, where it will be seen that the average drift was −0.4 Mc/s². The spectra sometimes show the presence of a second harmonic, but are often so complex as to preclude such identifications. Thirty-four slow drift bursts were recorded over the period 1956 October 1—1957 September 30; their duration was 175 min, which corresponds to a temporal probability of occurrence of 1 in 1400.

The characteristic velocity of the solar disturbances which give rise to these slow bursts may be deduced from their rate of change of frequency. This velocity is of the order of 1000 km/s and as it corresponds to that of the so-called "auroral corpuscular streams" it has long been held that the slow bursts are caused by the passage of the streams through the solar atmosphere. Alternatively, the bursts may be caused by acoustic shock waves resulting from explosions in the lower atmosphere. These waves, being propagated outward at some small multiple of the thermal velocity of the protons, would also be traveling at a velocity of the order of 1000 km/s.

Fast-Drift Bursts

Fast bursts, a very commonly occurring phenomenon, have durations of a few seconds and show exceedingly rapid drifts toward lower frequencies. They typically occur in groups of 3 to 10 with a total duration of less than 60 sec, as illustrated in Fig. 4. During the period 1956 October 1 to 1957 September 30, more than 14,000 fast bursts were recorded. No definite correlation has previously been found be-

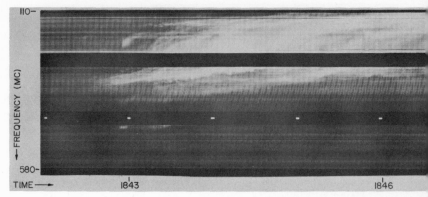

Fig. 3. Slow-dr**

tween the bursts and solar optical phenomena. However, a comparison of the first 10 months' observations at Fort Davis with the optical
data shows the following: 1) during periods when optical observations
were being made (approximately 50 percent of the radio observing
time) 50 percent of the burst groups were associated with flares; 2)
the positions of these flares were approximately uniformly distributed
across the solar disk, and 3) the reverse correlation, that of 1200 flares
with the radio bursts, was only 25 percent. Details of this work will
be published elsewhere.

The observations have also revealed a new type of fast burst which
appears on the records in the form of an inverted "U."[6] These bursts
show a rapid decrease in frequency followed by an increase, and have
a duration of a few seconds.

The source velocity which would be deduced from the slope of the
fast bursts is of the order of 100,000 km/s. It has therefore been suggested that the bursts are caused by the outward passage of solar
cosmic rays.[7] Alternatively it has been proposed that electron plasma
shock waves may provide the mechanism.[8] On this latter hypothesis
the "U" bursts would be caused by such waves being guided around
a magnetic field of force.

Continuum

The continuum radiation is a steady enhancement of the background level over a wide band of the spectrum, and, as described
earlier, is often associated with noise storms. At times, however, an
extremely intense form of continuum radiation is observed covering
a frequency band of more than 300 Mc/s. It often occurs after great
outbursts, and may last 10–300 min. This radiation probably corresponds to that described by Boischot,[9] from single-frequency observa-

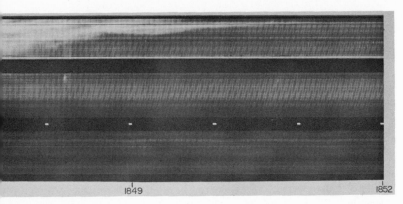

gust 2, 1843–51 U. T.

tions at 169 Mc/s, as "Type IV" radiation. Boischot and Denisse[10] have suggested that it may be caused by the synchrotron mechanism.

———————

The work described in this paper forms part of a program of research conducted under a contract with the Geophysics Research Directorate of the United States Air Force. The authors are indebted to S. J. Goldstein for assistance with the initial observations and gratefully acknowledge the help and cooperation of Dr. H. Jasik, M. Lebenbaum, and J. Goodman in the construction of the equipment. They also wish to thank Prof. D. H. Menzel, of the Harvard College Observatory, and Dr. J. Evans and Dr. E. Dennison, of the Sacramento Peak Observatory, for invaluable help in establishing the Station.

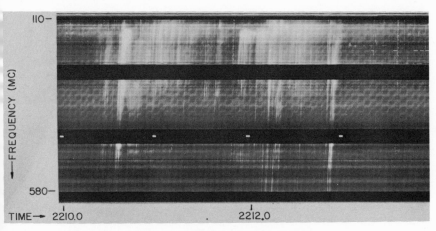

Fig. 4. Fast-drift bursts: 1958 July 27, 2210–13 U. T.

REFERENCES

1. This paper originally appeared in *Proc. I. R. E. 46,* 142 (1958), and is republished here by permission of the Institute. The original text has been slightly modified, and new illustrations have been substituted.
2. J. P. Wild and L. L. McCready, *Australian J. Sci. Research A 3,* 387 (1950).
3. J. P. Wild, J. D. Murray, and W. C. Rowe, *Australian J. Phys. 7,* 439 (1954).
4. H. Jasik, *Proc. I. R. E. 46,* 135 (1958).
5. J. Goodman and M. T. Lebenbaum, *Proc. I. R. E. 46,* 132 (1958).
6. A. Maxwell and G. Swarup, *Nature 181,* 36 (1958).
7. J. P. Wild, J. A. Roberts, and J. D. Murray, *Nature 173,* 532 (1954).
8. M. Krook, Unpublished.
9. A. Boischot, *Compt. Rend. 244,* 1326 (1957).
10. A. Boischot and J. F. Denisse, *Compt. Rend. 245,* 2194 (1957).

8

Natural Background Noise at Very Low Frequencies

JULES AARONS

To describe the term "very low frequencies" more exactly is to define the sphere of interest of this review paper. P. Goldberg[1] has suggested an apportionment of the spectrum as follows: 3–30 c/s, extra-low frequency; 30–300 c/s, super-low frequency; and 300–3000 c/s, ultra-low frequency.

We have modified these somewhat so that we discuss the range 0.3 to 30 c/s, calling this "extra-low frequency." We have included an additional spectral region which starts at 0.003 c/s and extends to 0.3 c/s; this we term "short-period micropulsations."

The Earth's Magnetic Field

While we shall not deal with the very-long-term changes in the earth's magnetic field, it might be well to mention the earth's "permanent" magnetic field.

The value of the total magnetic field F of the earth at geomagnetic colatitude θ is

$$F = F_o(1 + 3\cos^2\theta)^{1/2}, \tag{1}$$

where F_o is the equatorial intensity, approximately 0.3 gauss. Thus the steady-state magnetic field intensity is rather large, and is directed horizontally at the geomagnetic equator and vertically at the geomagnetic poles.

Perhaps the most prolonged irregular fluctuation of the earth's magnetic field is the sudden magnetic storm. Originating in the action of the corpuscular streams from radio-noisy sunspots or from solar flares on the ionosphere, these storms can last for a period of many days. However, even with the very large storms, the variational com-

111

ponent rarely amounts to more than 1000γ ($1\gamma = 10^{-5}$ gauss) at middle latitudes, or $\frac{1}{50}$ of the earth's field. The usual diurnal variations are of the order of 100γ. The magnetic storms may last from several hours to several days.

In 1859 the first observations of a large chromospheric eruption and its geophysical effects were made. An aurora associated with the solar flare was visible in England, where magnetic fluctuations of periods of 30 sec to 5 min were observed on the magnetometer trace by Balfour Stewart.[2] It should be noted that, with normal observatory magnetometers, the damping of the instruments, the sensitivity of the equipment, and the slow recording speeds limited the shortest period observable to the order of 30 sec.

Following the early work of Balfour Stewart, others have observed giant micropulsations with standard magnetometers during magnetic storms.[3] The Rolf oscillations are present for periods varying from many minutes to a few hours. During periods of low sunspot number, little evidence of the giant micropulsations is seen.

The variations in the earth's magnetic field can be determined by measuring the excursion from a normal value or by recording the changes directly. The normal magnetometer measures the fixed field plus the changes, while a loop and electronic amplifier are sensitive only to the rate of change of the earth's magnetic field.

The induced emf in a plane loop in a magnetic field is

$$e = -c\frac{d}{dt}\int B\,ds. \tag{2}$$

Here B is the magnetic field, namely, $(B_o + b\cos\omega t)$, where $\omega = 2\pi f$; hence $dB/dt = -\omega b\sin\omega t$. Therefore,

$$e = c\omega b\sin\omega t \times \text{turns} \times \text{area}. \tag{3}$$

Thus the loop–amplifier combination records only the rate of change of magnetic field rather than the standard magnetometer's total field.

To produce a signal amplitude of 1 μv the magnetic field must change 1γ/sec for a loop of 10^3 m^2 of area. In Eq. (3) we note the presence of the ωB term. Thus for field intensity in microvolts per meter, the product ωB is constant. To standardize results we confine our field-intensity levels to those which are contained within a bandwidth of 1 c/s. With white noise, which is uniform in amplitude across the frequency spectrum, the amplitude increases with the square root of the bandwidth. Therefore, the field intensity should be measured in microvolts per meter per cycle per second. To change from gamma (10^{-5} gauss) to microvolts per meter, the gamma values are multiplied by 3×10^5.

The orientation of the axis of the coils used determines the component of the earth's field to be measured, since only the variation of the component in the direction of the coil axis is recorded. In many studies of the fast fluctuations of the earth's field, only the horizontal component has been recorded at low and middle latitudes.

The Four Spectrum Components

Short-period Micropulsations: 330–3.3 sec

The general characteristics of fluctuations in this field have been described by Holmberg.[4] A magnetometer with a threshold of $10^{-2}\gamma$ compared to the usual 1-γ sensitivity was set into operation during the 1932–33 Polar Year. The unit was capable of recording fluctuations with periods ranging from 100 sec to 1 sec.

During the day the continuous amplitude of the fluctuations was high. The periods were relatively long with two predominant—one at 20 sec and one at 70 sec. The maximum level of this daylight phenomenon was of the order of 20γ. After sunset a comparative quiet reigned. A nighttime phenomenon was a train of sine waves, each group of waves lasting 5 to 10 min. The periods of the oscillating components within the train, or burst, varied between 30 and 130 sec. The maximum number of these bursts, or trains, was received between 2200 and 2300 hours.

In recent measurements encompassing this region of the spectrum, Duffus, Nasmyth, Shand, and Wright[5] at the Pacific Naval Laboratory of Canada verified the presence of nighttime bursts. Discontinuous signals were found with a maximum centering about local midnight. In addition to the normal diurnal variations, geomagnetic disturbances were detected at levels clearly above the continuous background. Descending tones similar to whistlers were recorded on magnetic tape with frequency components of several seconds. During periods of magnetic quiet, the "whistles," lasting for many minutes, were usually of the falling type. During geomagnetic disturbances, both rising and falling whistles were detected.

Both Rolf[3] and Suckdorff[6] had already noted rapid fluctuations at northern latitudes. In 1932 Suckdorff observed fast micropulsations with periods of the order of 2–3 sec. These lasted for 1 hr, several hours, or sometimes even for 24 hr continuously, after which a considerable period passed before the normal magnetic levels appeared.

Lubiger[7] found that the frequency of pulsations recorded during the 4 hr centered around midnight was higher than that for the 4 hr centered at noon. Shorter pulsations (less than 90 sec) were more heavily concentrated at night than the longer ones. Duffus and Shand[8] have found the long-period trains peak around midnight, but not the over-all amplitude.

The correlation of signal level in this spectral region with magnetic storms indicates that the source of fluctuation is magnetic. The amplitude of the energy reaches its peak during the day, while the night is punctuated with short, signallike bursts. Thus short-period, continuous high-amplitude signals are recorded during the daylight hours. The night shows groups of signals of smaller amplitude but consisting of longer-period components.

Extra-Low Frequency: 0.3–30 c/s.

Perhaps the most controversial portion of the spectrum under consideration in this study lies in this region. The fluctuation amplitudes are small—of the order of 2 to $20 \times 10^{-3}\gamma(c/s)^{-1}$. The question under dispute is whether the predominant origins of the signals are fluctuations of the earth's magnetic field or atmospherics.

We shall first present the data purporting to show the importance of the earth's magnetic fluctuations. In experiments with the coil and amplifier technique, Aarons[9] reported high-level signals during some magnetic storms. Maple,[10] working at frequencies of 3–45 c/s in arctic regions, has detected large fluctuations several hours before magnetic storms. At low latitudes this phenomenon has also been detected. The proponents of the magnetic-fluctuation hypothesis point to the increase in amplitude accompanying magnetic disturbances and to the increase in amplitude at high latitudes relative to middle-latitude experiments. Wilcox and Maple[11] found that at frequencies below 1 c/s the geomagnetic fluctuations at White Oak, Maryland, were more definitely associated with ionospheric sources. In this band of frequencies the received signals were considerably greater at the auroral latitudes than those detected at White Oak.

Northern-latitude atmospherics are smaller in amplitude and occur less frequently than those recorded at middle latitudes. However, the signal levels detected by Wilcox and Maple increased with increasing latitude. Aarons and Henissart[12] have shown a positive correlation between 0.5–20-c/s noise and the magnetic K index.

The point of view taken by several observers—Goldberg,[13] working between 1 and 150 c/s, and Schumann and Konig,[14] working between 1 and 25 c/s, is that the lower atmosphere is the basic source of the bursts observed. Goldberg finds a maximum of activity at 1500 hours local mean time with a second minor maximum at 2400 hours. He ascribes the fluctuations largely to lightning activity. Other observers have found local atmospherics activity at a maximum in the afternoon, particularly in the geographic location where Goldberg conducted his experiments, while the magnetic fluctuations reach their peak around local midnight. Perhaps these data should be divided into two components, magnetic and atmospheric, thus distinguishing the origins of the two maxima.

Schumann and Konig, on the other hand, found peaks in the afternoon (1600–1800 hours in the summer and 1300–1500 hours in the winter) and ascribed the activity to temperature inversions at low altitudes. They recorded atmospherics but stated that the main component of the signals was not of this type. They noted the presence of wave trains.

We would now like to note, at least in order of magnitude, the amplitude level of these noise background signals. At the long-period end of the spectrum (periods of the order of 100 sec), fluctuations of 20γ are observed during magnetic disturbances.[4, 10] At 10 c/s, signals of 2 to 5 $\times$ $10^{-3}\gamma$ c/s have been recorded.[9, 13, 15] Frequently, at 10 c/s, sporadic bursts would increase the signal amplitudes by a factor of 10. The amplitude varies inversely with some power of the frequency. It is very difficult to ascertain a more precise dependence, for many reasons. First, the data come from experiments which vary widely in experimental technique and calibration methods. The differences in latitude and longitude of the various experiments constitute a second factor that makes normalization of results difficult. Geomagnetic latitude is important for its effect on the magnetic components of the signal, while geographic latitude and longitude are important for their relation to thunderstorms and for their effect on the propagation of the atmospherics.

Another factor to be considered in the fluctuations of short-period micropulsations and extra-low frequencies is the semilocal nature which Selzer[16] has found to be a characteristic of certain fluctuations of the earth's magnetic field. Figure 1, from Selzer, shows that the large oscillations observed in the magnetograms of western European observatories were not observed in American observatories. The separation of stations at which the oscillations were correlated was of the order of hundreds of kilometers. Correlated signals were recorded at stations separated by 3000 km by the scientists[5] at the Pacific National Laboratory and by Aarons and Henissart[12] during some magnetic storms, but not in all.

In commenting on the burst maximum during the evening hours, Ashour and Price[17] have suggested that the day-night contrast is due to the electromagnetic shielding properties of the ionosphere. Diurnal ionization produces a conducting layer which is 4 times greater in the F-layer and fifteen times greater in the E-layer during the day than at night. If the impulse type of micropulsation is caused by shocking

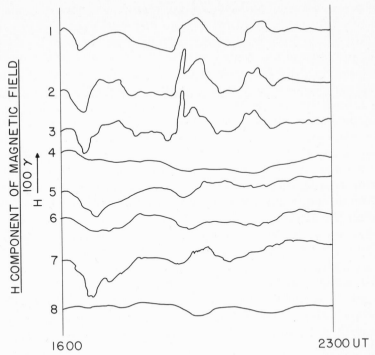

Fig. 1. Recordings of horizontal component of earth's magnetic field (Selzer) for: 1, Tamanrasset; 2, Chambon-la-Forêt; 3, Nantes; 4, San Juan; 5, Cheltenham; 6, Tucson; 7, Sitka; 8, Honolulu.

the ionospheric currents, the large daytime currents will shield ground detectors from small changes. The nighttime pulsations are a relatively large part of the total nighttime ionospheric currents.

The foregoing hypothesis is in good agreement with Huancayo data. At that equatorial observatory, irregular pulsations of periods greater than 1 min are more prevalent during the day. Huancayo does not have the large differences in ionospheric currents between day and night that other observatories farther north or south show. It has more constant overhead currents, day and night. The shielding effect of the daytime ionosphere is nullified at Huancayo, and a larger number of micropulsations of irregular nature are received during the day.

Super-Low Frequencies: 30–300 c/s

The 30–300-c/s portion of the spectrum, containing the power-line frequency and its harmonics, is extremely difficult to instrument. Various experimenters have used battery operation and complex techniques to eliminate both the power-line signal and its transients.

The propagation of power-line energy and the effect of transients make it difficult to reduce the background level to numerical values. At sites remote from power, long-term data are extremely tedious to take, and most of the studies outlined have been of relatively short duration. There is good agreement with respect to the source of energy —that is, radiation from atmospherics, radiation which undergoes only a small amount of attenuation.

Using both a loop and a vertical whip antenna, Holzer and Deal[18] have taken a series of measurements which correlated signal intensity in the region 30–110 c/s with air-earth current density near the sea. The main signal found was roughly proportional to the number of storms in progress over the surface of the earth. The origin of the electromagnetic signals was ascribed to propagation from world-wide lightning discharges.

The low absorption of energy in the region 20–200 c/s has also been shown by Aarons[20] in New Mexico. In the sweep-frequency measurements made with a narrow-band filter, the signal level was found to be a maximum in the spectral region 20–200 c/s. In the tracings of the recordings shown in Fig. 2, only a small diurnal variation is seen relative to the signals recorded in the spectral region 200–900 c/s. The latter signals showed great fluctuations between day and night values.

Many of the data in this band have been taken at middle latitudes where atmospherics are predominant. Future work centering on auroral latitude studies may show the contribution of the magnetic component.

Ultra-Low Frequencies: 300–3000 c/s

The 300–3000 c/s spectral region is characterized by high attenuation during the day, by a background level which decreases with increasing latitude, and by the signal varying inversely with frequency. Utilizing atmospherics originating in European storm centers and in discharges over the Atlantic, Chapman and Macario[19] have found that the minimum attenuation over both land and sea is at the 100-c/s end of the frequency scale (100 c/s–10 kc/s was the spectrum analyzed). Figure 3 gives a large-scale picture of attenuation in the audio-frequency spectrum. Although land and sea attenuation as a function of frequency were similar in outline, a much higher attenuation took place between 1 kc/s and 4 kc/s over land than over sea. Effectively the earth-ionosphere acts as a band-pass filter in the region of 30–300 c/s and a band-rejection filter around 2.5 kc/s. Figure 4, from Wilcox and Maple, gives the magnetic-field fluctuations at two sites as a function of frequency.

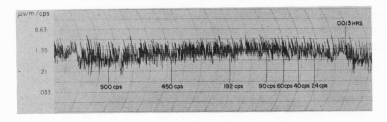

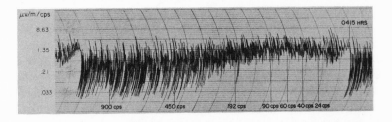

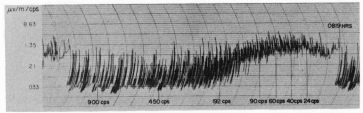

Fig. 2. Sweeps recorded at 10–900 c/s, on a vertical antenna, at 0013, 0415, and 0819 local time on 1 September 1955. A time period of 14 min, 20 sec is shown.

Several papers[21] have recently appeared with excellent evidence that there are emissions of energy from the aurora or associated with micropulsations. Signals have been recorded in the region above 300 c/s, extending into the 10 kc/s portion of the spectrum. The background level, however, is still set in all likelihood by atmospherics and the emissions are enhancements of this level, evident during ionospherically disturbed periods.

Low-Frequency Phenomena

Many investigators have found trains of signals. These lasted for several seconds to several minutes. They have been predominantly sinusoidal or narrow-band, and have changed in period during a single train. They are frequently grouped together, start abruptly, display the variation in period during their single grouping, and then decay. Activity of 1–5-sec duration has been recorded by Aarons[22] and is illustrated in Fig. 5. The individual train lasted from 10 to 30 sec.

Very-low-frequency whistles have been recorded by the Pacific National Laboratory[5] with frequency components of 2–12 sec, the whistles lasting from 1 to 4 min. Holmberg[4] and Schumann and Konig[14] have also noted the presence of these oscillations. So far, these phenomena have not been definitely correlated with other atmospheric or geomagnetic phenomena.

Another phenomenon which will bear investigating is that of absorption or radiation of signals at the gyro-frequency of ions in the earth's atmosphere. While a small number of data are available,[20] observations are relatively sketchy. Detection of these gyro radiations would be extremely valuable for upper atmosphere research.

Conclusion

Although at the time of writing the portions of the spectrum and their characteristics are reasonably well delineated, much work has

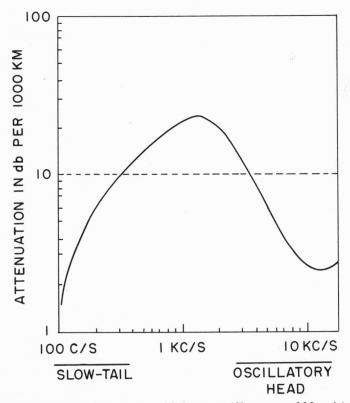

Fig. 3. Variation of attenuation with frequency (Chapman and Macario).

to be done on long-term characteristics of the signals. The fluctuation component data lack the consistency of recording and of calibration of the observatory magnetometers. Research procedure must be changed to that of normal observations. Methods of calibrating equipment and analyzing data are being standardized by various groups so that networks of stations can be set up throughout the world.

Discussion

P. A. Goldberg, *Boeing:* I should like to add two comments to Dr. Aarons' excellent summary. One, which should intrigue radio propagation people and radio engineers, has to do with the kind of attenuation that you have at these very low frequencies of about 100 c/s. As you can see, from the study by Chapman and Macario which was done for measurements on lightning strokes at distances of about

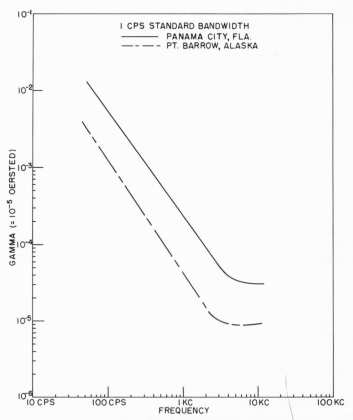

Fig. 4. Frequency spectra of magnetic fluctuations. Horizontal component, fall 1952 (Wilcox and Maple). Average of "steady background" signals observed in 10-min scaling intervals.

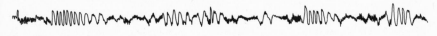

Fig. 5. Trains of signals (minute markers shown).

1000 km, and from approaches such as Dr. Aarons and I have also used which are concerned with thunderstorm centers at distances of 5000 to 10,000 km, the attenuation is extremely low, less than 1 db/1000 km. You can monitor total worldwide thunderstorm activity this way. Holzer of U.C.L.A. is now concentrating on this approach.

My second point concerns what you might call the transition region of the spectrum. It looks now, although nothing is very definite, as though the worldwide thunderstorm activity propagates extremely well down to about 10 c/s. It then starts to diminish, because the earth's ionosphere is no longer exactly a waveguide, since the distance around the earth is one wavelength at 10 c/s. Below 10 c/s the other phenomena, presumably due to fluctuations in ionospheric currents, appear very strongly. Thus, in a spectral plot for the region 1 to 100 c/s, we have intensely high spectral density at one cycle falling off uniformly up to about 100 c/s.

REFERENCES

1. P. A. Goldberg, private communication, July 24, 1958.
2. Balfour Stewart, *Encyclopedia Britannica,* (9th ed., 1882), p. 36.
3. B. Rolf, *Terrestrial Mag. Atm. Elec. 36,* 7 (1931).
4. E. R. R. Holmberg, *Monthly Notices Roy. Astron. Soc.* (Geophys. Suppl.) *6,* 35 (1953).
5. H. J. Duffus, P. W. Nasmyth, J. A. Shand, and C. S. Wright, *Nature 181,* 1258 (1958).
6. E. Suckdorff, *Terrestrial Mag. Atm. Elec. 52,* 147 (1947).
7. F. Lubiger, dissertation, University of Göttingen, 1924.
8. H. J. Duffus and J. A. Shand, *Can. J. Phys. 36,* 508 (1958).
9. J. Aarons, *Proc. Am. Acad. Arts Sci. 79,* 266 (1951).
10. E. Maple, Abstract, Amer. Geophys. Union Meeting (May 4–6, 1953), p. 334.
11. J. B. Wilcox, and E. Maple, Navord Report 4009, U. S. Naval Ordnance Laboratory, White Oak, Maryland, July 9, 1957.
12. J. Aarons and M. Henissart, *Nature 172,* 682 (1953).
13. P. A. Goldberg, *Nature 177,* 1219 (1956).
14. W. O. Schumann and H. Konig, Die Naturwiss. *8,* 183 (1954).
15. H. F. Willis, *Nature 161,* 887 (1948).
16. E. Selzer, *Ann. Geophys. 8,* 275 (1952).

17. A. A. Ashour and A. T. Price, *Proc. Roy. Soc. A 195*, 198 (1948).

18. R. E. Holzer and O. E. Deal, *Nature 179*, 536 (1956).

19. F. W. Chapman and R. C. V. Macario, *Nature 177*, 930 (1956).

20. J. Aarons, *J. Geophys. Research 61*, 647 (1956).

21. J. M. Watts, *J. Geophys. Research 62*, 199 (1957); A. Egeland, Report No. 6, Institute of Theoretical Astrophysics, Blindeen Oslo (1959); R. M. Gallet, *Proc. I. R. E. 47*, 211 (1959); G. R. A. Ellis, *J. Planetary and Space Sci. 1*, 253 (1959); E. Ungstrup, *Nature 184*, 806 (1959); J. Aarons, G. Gustafsson, and A. Egeland, *Nature 185*, 148 (1960).

22. J. Aarons, doctoral thesis, University of Paris, June 1954.

9

Solar Whistlers

THOMAS GOLD and DONALD H. MENZEL

We wish to discuss a story somewhat related to the phenomenon of terrestrial whistlers which we have recently tried to put together. Nowadays, good evidence from cosmic rays clearly indicates the presence of substantial magnetic fields, of the order of 10^{-4} gauss, in the space between the sun and the earth. It is necessary to suppose that these fields frequently are elongated—very well combed—so that they stretch radially from the sun to the earth, or beyond. We are not sure about the configuration beyond the earth, but this kind of field commonly occurs in the space between the sun and earth. We infer the existence of this radial field from the fact that cosmic-ray particles of undoubted solar origin have arrived at the earth on several occasions only a short interval after their presumed origin in a visible solar flare. Such particles, whose initial direction, before deflection by the earth's field, was from the sun, suffered only a short time delay. By contrast, particles approaching the earth from a direction other than that of the sun experienced considerable time delays. A magnetic field furnishes the simplest mechanism to account for these delays. A magnetic field strong enough to deflect the particles must have been present, but since they arrived with a minimum delay from approximately the direction of the sun, we conclude that this field was arranged mostly in the earth-sun direction. For that reason, at the URSI conference in Boulder (August 1957), one of us (Gold) suggested that such a field stretching from the sun to the earth might cause "whistlers," analogous to those associated with the outer field of the earth. We have now looked into this possibility, and a number of interesting points have come to light.

The range of expected frequencies would of course be very much lower than for terrestrial whistlers because the fields are only some

10^{-4} gauss, considerably weaker than that of the earth. This condition limits the highest frequency that can be present in the phenomenon. Secondly, one would expect the dispersion to be very much greater, chiefly because of the extremely long path between the sun and the earth in comparison with the short paths for ordinary whistlers. For this reason one would expect frequencies roughly of the order of 1 c/s, or indeed anything between about 100 c and 0.1 c/s.

What sort of observed phenomena occur in that frequency range? Periodic fluctuations, usually called "pulsations," in the earth's magnetic field are well known; the largest of them occur in the polar regions. The extreme regularity of the phenomena has been puzzling. Why, for an interval as great as half an hour, should a regular periodic signal arrive? Figure 1 shows such a signal, with other disturbances unavoidably superposed. The oscillations are unmistakably regular, although the amplitude does not stay quite constant. These pulsations could conceivably originate in one of two alternate ways: in some sort of oscillating device of high Q close to the earth, or as the dispersed spectrum of an impulsive burst at great distances. If the second explanation is correct, the frequency would have to change gradually, according to the group velocity of the various frequencies present in the original pulse. The frequency change could be either up or down, as for terrestrial whistlers. A dynamic frequency analysis gives information about the origin of the oscillations.

Figure 2 shows a frequency analysis, kindly furnished us by Dr. H. J. Duffus and Sir Charles Wright of Canada. They record on magnetic tape, at a very low speed, fluctuations in the earth's magnetic field. Then, in order to have a rough and quick frequency analysis, they play the tape back a thousand times faster. The ear then serves as a good frequency analyzer. Figure 2 displays an actual frequency analysis from such a tape on a particular occasion when a rising "pitch" changed from 2 c/s to 3 c/s in about 15 min. There are many

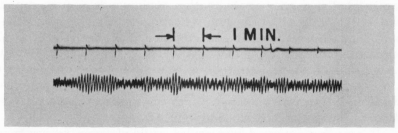

Fig. 1. Magnetogram showing nearly periodic oscillations. (Courtesy of H. J. Duffus and C. Wright.)

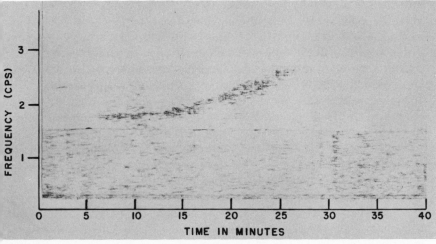

Fig. 2. Frequency analysis of nearly periodic oscillations, showing gradual drift in frequency—rising whistle. (Courtesy of H. J. Duffus and C. Wright.)

such instances. Both rising and falling "whistles" are observed, as is also the case with the earth's whistlers. Independently of our suggestion, they had commented that the signals, speeded up by this factor of 1000, closely resembled terrestrial whistlers.

Let us consider the orders of magnitude involved. The dispersion of a whistler is directly proportional to the path length and the square root of the electron density, and inversely proportional to the square root of the magnetic field. For the earth-sun space with an electron density of $100/cm^3$ and a magnetic field strength of 10^{-4} gauss, the delay for a 1-c/s wave would be of the order of 20 days. It is known that motions frequently occur that transport gas masses from the sun to the earth in a fraction of this time. The expectation is therefore that an undisturbed "whistler" train is a most unlikely phenomenon, and that the frequency received is commonly greatly affected by the Doppler shift appropriate to a moving medium of high refractive index. In a really fast-moving stream it is even appropriate to consider the wave as being almost locked up in the gas and bodily transported. The observed changes in frequency would also be expected to relate mostly to changes in the Doppler shift.

The highest frequency expected without Doppler shift for a field strength of 10^{-4} gauss would be of the order of 100 c/s. The delay time for this wave for the sun-earth propagation would be only 2 days for the same assumed values of conditions as before.

If this interpretation is correct, the phenomenon of solar whistlers provides a way of finding out more about the space between the sun and the earth. It may be possible to find correlations with those features of cosmic-ray phenomena that are dependent on a suitable magnetic field between the sun and the earth. We might then expect a decrease in the low-energy cosmic-ray flux at the earth to occur during a period of pronounced solar whistlers. There is also, of course, the possibility of finding an actual correlation with particular disturbances on the sun. We should expect to receive the whistler signals whenever a suitable propagation path exists, as an adequate amount of noise is probably constantly generated at the solar surface. The observed frequency-time curve, complicated by the Doppler effect of the moving medium, may prove to be a way of obtaining information about the condition and motion of the medium.

The propagation conditions in the extreme vicinity of the earth are a further difficulty. The lines of force of the earth's field do not in general connect with the large-scale field through which the signal is propagated. There has to be some transition from the one field to the other, and it is clear that in the dense medium close to the earth a signal in this frequency range can propagate only as an Alfvén wave. This coupling needs to be investigated, but it is immediately clear that it will be tighter to the polar lines of force than to the equatorial ones. The equator is shielded by all the overlying field, whereas the polar lines of force reach far out into a tenuous medium and come close to the lines of force along which the signal is traveling. The best place to look for solar-whistler phenomena would thus be the vicinity of the geomagnetic pole, and there are some indications already that the effects in this frequency range are greatly enhanced in polar regions.

Discussion

R. S. Lawrence, *National Bureau of Standards:* Is there any other evidence for this combing out of the magnetic field? And second, does this imply that streams of material, leaving the sun, cause this combing?

T. Gold: The evidence for this combing out of the magnetic field is the following: First, any disturbance on the surface of the sun that throws out material must set up an elongated field if the gas originally had lines of force running through it that connected into the solar surface. This connection will remain unbroken—the conductivity of these gases is high enough to assure that—and therefore, if such gas reaches the earth, lines of force must exist that go from this gas back to the sun. One might debate whether the field is likely to be as straight as I have been suggesting, or whether some kind of turbulent phenomenon would put it into more complex shape. I think, in fact, that turbulence cannot occur on this scale because of the long mean free path in the medium.

Second, the cosmic rays that are produced in a solar outburst have been seen to arrive first from the direction of the sun, and later from a variety of other directions. To have the wide-angle distribution later, one must conclude that a field existed in the space; but to have initially the direction from the sun, it is necessary to suppose that the field was in the earth-sun direction and that these were the particles with very steep pitch angles in the field. Much later in the phenomenon the whole magnetic bulge would be filled with all kinds of orbits of particles, and the earth would be bombarded more or less isotropically. This indeed seems to have been the case.

R. A. Helliwell, *Stanford University:* If you apply the simple formula that I believe you were using on the dispersion, it looks as if very long time delays would be involved. The time for light propagation is of the order of 8 min from sun to earth, and the reduction in group velocity for the figures you are using is of the order of 10^{-3}, so the time delay for the solar whistlers would be of the order of 8000 min.

Gold: So far as I know, long delay times are demanded in any case without any estimate of the actual conditions in space, just by the observed value of the dispersion. From this we can surely extrapolate the delay time, and it is true that this turns out to be very long. The chance of obtaining a signal that has traveled from the sun to the earth and curved back to the sun and then returned to the earth is not very great. The motion of the earth and the motion of the gases will in general tend to interrupt a wave train much more quickly than that, and we can really expect only short bits of interrupted wave train. It is of interest that there appears to be a positive correlation between the whistler propagation and periods following magnetic storms.

E. Maple, *AFCRC:* The rapid periodic geomagnetic fluctuations at frequencies above 1 c/s may fit the proposed theory; few data are so far available. For periods longer than 1 sec, however, the geomagnetic pulsations do not fit the theory, since the observed periods vary with local time and are sometimes essentially constant over a period of several hours. For example, there have been some observations of 30-sec periods, which do not vary more than 10 percent over an interval of 3 or 4 hrs; but the signal would show an oscillation, it would die down and another one start, while its period would stay within about 10 percent for hours. If this is true, I think the solar-whistler phenomenon will have to be restricted to frequencies above about 1 c/s.

Gold: I am not sure how far down the frequency scale this phenom-

enon can reach, but at the very low frequencies the expected dispersion would be very great. We would thus expect to see signals of
remarkably constant frequency, quite in agreement with the observations you mention.

O. E. H. Rydbeck, *Chalmers University, Sweden:* What is the
effective wavelength of this electromagnetic radiation, say at 2 c/s?

Gold: 150 km.

Rydbeck: Is it necessary to assume coupling to Alfvén waves in
order to explain the received electromagnetic field at the earth's
surface?

Gold: Yes, the ionosphere cannot transmit the wave significantly
in any other mode of propagation.

Helliwell: As the energy approaches the earth, the gyro-frequency
increases, and this would increase the wavelength, by an order of
magnitude, from 150 to 1500 km or so.

Gold: Yes, but even so it will be an Alfvén wave that actually
comes through the ionosphere, because I do not think that at a frequency of 1 c/s there is any other mode of propagation in these
regions. There will be a smooth transformation from the whistler mode
into an Alfvén wave as the density and field strength increase.

10

Noise of Planetary Origin

B. F. BURKE

Planetary radio noise, like solar radio noise, can be classified as thermal or nonthermal, depending on its characteristics. The easiest to understand, though by no means the easiest to detect, is the thermal component, which is simply the long-wavelength tail of the Planck radiation from a hot body. The nonthermal radiation, on the other hand, is characterized by unusual intensity, and can be very easy to detect but very difficult to understand. We shall first consider the characteristics of the thermal component.

The observations of thermal radiation from the planets at the present time have been made at centimeter wavelengths, using microwave radiometers in conjunction with large steerable paraboloids. Indeed, the thermal origin of the radiation necessitates the use of as short a wavelength as possible, for the incident energy flux falls rapidly with increasing wavelength, while the background noise from the galaxy rises, thus making detection more difficult. Because of these factors, it is unlikely that the planetary thermal radiation can be studied at wavelengths greater than a few tens of centimeters, although the same reasons certainly favor the extension of observations at the other end of the radio spectrum, into the millimeter domain.

The observations reported so far on the planets can only be considered to represent a beginning, for the rapidly improving techniques of microwave radiometry will certainly provide more detailed data in the near future. Venus, Mars, Jupiter, and Saturn have all been detected, by C. H. Mayer et al.[1] and by Drake and Ewen.[2] The deduced apparent black-body temperature of Venus, as determined by Mayer et al.[1] is sufficiently accurate to be of astrophysical interest, and can serve as an example of how the microwave measurements can be interpreted. The observed apparent black-body temperature at a wavelength of 3 cm appeared to change slightly with the phase of Venus, but averaged $560° \pm 73°K$ at inferior conjunction, corresponding to an energy flux at the earth of 10^{-24} w m^{-2} sec. A sample record in which the planet was allowed to drift through the beam of the NRL 50-ft dish is shown in Fig. 1. Averaging a large number of such records was required to obtain improved accuracy, although more recent developments in receivers allow much improvement in signal-to-noise ratio. In order to interpret this observation, we consider a simple model based on the rather meager optical data available. Venus possesses an atmosphere of which the only reliably identified constituent is CO_2. The amount of O_2 and H_2O, if present, is less than 5 percent of the amount in the earth's atmosphere, according to Dunham.[3] There are dense, highly reflecting (in visible light) clouds of unknown composition which prevent good visual observations of the presumably solid surface. If the atmosphere is transparent to the radio wavelength used for observation, all the radiation originates from the surface of the planet, and if the surface is approximately a black body, the observed brightness temperature corresponds to the average physical temperature of the surface. The electrical conductivity and dielectric constant of the surface affect the depth of penetration of the radio waves, and hence the depth over which the effective temperature is averaged. By measuring the apparent temperature as a function of wavelength it might be possible to gain some knowledge of the variation of temperature with depth, since different wavelengths will average the temperature over different depths. This technique has been applied to the moon,[4] where the variation of temperature with phase, that is, variation with incident solar energy, provides still another variable which can be used in constructing models of the surface.

When an atmosphere is present, as is certainly the case with the planet Venus, absorption and emission processes in the atmosphere must be carefully considered. For example, in the extreme case of a nearly opaque absorbing layer at the observation wavelength, the observed radio temperature will refer to the temperature of that layer. A planet with a partially absorbing atmosphere will produce an observable radio temperature with contributions representing both the atmospheric temperature and the surface temperature. (An exact treatment of the problem requires, of course, the solution of the equation of transfer.) Two general classifications of absorption mechanism will be considered: transitions between discrete molecular energy levels and transitions between a continuum of states. Examples of the first class of absorption mechanism can be cited in our own atmosphere, for strong absorption occurs for wavelengths near 5 mm, because of transitions in molecular oxygen, while a weak absorption at 1.25 cm

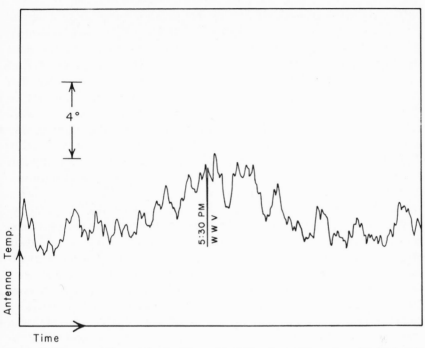

Fig. 1. Drift curve of Venus taken by Mayer, McCullough, and Sloanaker with NRL 50-ft paraboloid, at a wavelength of 3 cm. Peak deflection represents an antenna temperature of 3.5°K.

is attributable to water vapor. The planet Venus, however, has only one identified atmospheric component, CO_2, which has no absorption lines in the microwave region. The presence of other gases which might possess absorptions cannot be eliminated, but on the basis of existing evidence one must conclude that no known source of molecular absorption exists in the atmosphere of Venus. The second class, continuum absorption, is most effective in ionized media. The atmospheres of the planets are being bombarded constantly by ionizing radiations from the sun, resulting in the formation of an ionosphere at the top of the atmosphere. Here absorption is possible through either electron-ion collisions or electron-atom collisions. Some studies have been made of these effects in our own ionosphere, for example, by Mitra and Shain.[5] Near the critical frequency, strong absorption is observed, but at frequencies more than three times the critical frequency, absorption is negligible. An exception must be noted, however, in polar regions, where flare-associated absorptions have been observed at 75 cm, presumably caused by low-energy particle bombardment. Except for these exceptional effects, however, one may safely assume that continuum absorption in the earth's ionosphere is not of great importance at wavelengths shorter than 10 m. Considerable uncertainty is introduced when one transfers terrestrial ionospheric properties to Venus, for that planet certainly is subjected to more intense ionizing radiation. Nevertheless, if the ion densities were higher by a factor of 10^4, representing critical frequencies a hundred times those occurring in the terrestrial ionosphere, it would appear unlikely that these absorption processes would seriously modify centimeter wave radiation arising from lower levels.

The tentative conclusion may be made that the observed temperature probably represents a lower limit for the surface of the planet. The interesting physical implication of such a high temperature is the necessary absence of water on the surface of the planet. This conclusion would be strengthened, however, if the apparent temperatures were known as a function of frequency. A nearly constant or slowly varying apparent temperature would favor this interpretation, while the appearance of humps or dips, representing molecular absorptions, would indicate that the atmospheric contribution would have to be considered. The latter will certainly be the case for the major planets, with their extensive atmospheres.

Nonthermal planetary radiation presents a different problem altogether. On does not know in advance which objects in the sky will be strong sources of nonthermal radio noise, and little is known concerning the mechanisms responsible for producing such noise. At the present writing, only the planet Jupiter has been conclusively identified as a source of nonthermal radio noise, excluding the earth, which all radio astronomers concede is a copious source of nonthermal noise. The identification was made by Burke and Franklin[6] using the large Mills cross of the Carnegie Institution of Washington, although the noise signals were so intense that most later work has been done with much simpler equipment. Some of the salient characteristics of the nonthermal Jovian radiation are shown in Fig. 2. These records were made with two very simple interferometers, operated at a wavelength of 13.5 m (22.2 Mc/s); from them the polarization and energy flux of the incident radiation can be determined. Most of the observed radio noise takes the form of groups of intense bursts, the activity continuing for periods as long as 1 or 2 hr. Each burst is of the order of 1 sec in duration, although a second class of burst, of only a few milliseconds' duration, may exist. The peak energy flux per unit bandwidth often exceeds the flux of the most intense radio sources, maximum intensities of much greater than 10^{-21} wm^{-2} sec occurring commonly at this wavelength. If the entire planet were acting as a

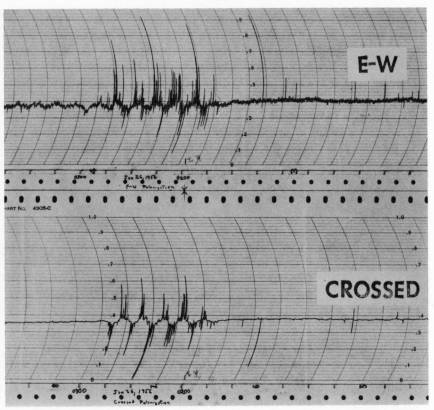

Fig. 2. 22.2-Mc/s interferometer records of noise bursts from Jupiter, obtained by Franklin and Burke at Carnegie Institution of Washington. Upper record was taken with interferometer having east-west polarization for both antennas, while lower record was taken with one antenna polarized east-west and the other antenna polarized north-south.

black body, a temperature well in excess of $10^{11\circ}K$ would be required, a value so high that a thermal origin for the noise is highly unlikely. A possibility that immediately suggests itself is thunderstorm activity, although the observed intensity, when compared with noise from lightning strokes on earth, is greater by a factor of order 10^9.[7] If electric discharges are responsible, they must be very different from those observed on the earth.

Further evidence for the unusual nature of the Jovian radio noise can be deduced from a comparison of the two records in Fig. 2, which were made simultaneously in order to measure polarization characteristics of the noise bursts. The upper record was taken with an interferometer both of whose elements were linearly polarized in an east-west direction, while the lower record was taken with an interferometer whose elements had crossed linear polarization, one east-west and one north-south, similar to the arrangement used for solar radio bursts by Little and Payne-Scott.[8] If the noise bursts were unpolarized, an interference pattern would appear on the upper record only, but by comparing relative phases and amplitudes of the two records, the degree of linear or circular polarization can be measured. In the example given, most, though not all, of the bursts are strongly circularly polarized, the polarization being nearly 100 percent in some cases. This characteristic reminds one of the radio-noise bursts observed from the active sun, which are well known to exhibit strong circular polarization much of the time.

Another peculiarity which distinguishes these nonthermal noise bursts from noise of thermal origin is the character of the spectrum observed. No confirmed observations of radio-noise bursts from Jupiter exist at wavelengths shorter than 10 m, while at longer wavelengths the noise bursts are so intense that they are easily observable with the simplest of equipment. Franklin[9] noted that a comparison of records taken at frequencies of 18, 22, and 27 Mc/s showed that noise could be observed at one frequency and not at another even though, at one time or another, Jovian radio noise was observable at all three frequencies. Gallet and Bowles, in unpublished work, have compared the noise observed at 18 and 20 Mc/s, with almost identical receiving equipment, and concluded that at the lower frequency, that is, longer wavelength, the noise bursts were more frequently observed and definitely more intense.

The existence of at least one localized source of noise on the planet was demonstrated by Shain,[10] who compared the occurrence of noise in his prediscovery observations of 1950–51 with the rotational period of the planet. In visible light, one can observe only the dense clouds near the top of the atmosphere of Jupiter, which exhibit a variety of rotational periods depending on their latitude. Figure 3 shows a comparison of the occurrence of radio noise at 18.3 Mc/s plotted as a function of the central meridian in two longitude systems, System I approximating the rotational period of the equatorial belt, and System II approximating the average rotational period of the clouds at higher latitudes. A systematic drift is evident in the System I longitudes, while the similar graph in the System II coordinates shows that there was definitely a higher probability of observing radio noise when a certain range of longitudes was near the central meridian. Consequently, the most active radio region on the planet Jupiter during late 1951 rotated at approximately the period of System II, which is by convention set at $9^h 55^m 40^s.632$.

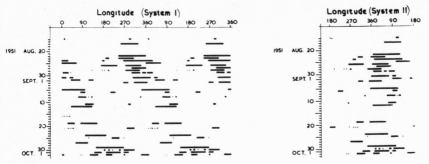

Fig. 3. Occurrences of radio noise from Jupiter observed at 18.3 Mc/s in Sydney and shown as a function of the longitude of central meridian in System I and System II (Shain[10]).

The lifetime of the active radio region is a question of immediate interest, and some information concerning it is provided in Fig. 4, which is a representation of all the published data on radio-noise occurrences on the planet Jupiter, and includes unpublished data as well. The histograms represent the number of observations of radio noise as a function of longitude in System II, from the following sources:

January 1951:   18.3-Mc/s observations of Shain;[11] only the position of the most intense region was given.

August 1951:   18.3-Mc/s observations of Shain.[11]

June 1954:   22.2-Mc/s observations of F. G. Smith and Burke (unpublished).

March 1955:   22.2-Mc/s observations of Burke and Franklin;[6, 7] smoothed by 40° in longitude since only point observations were taken.

December 1955:   22.2-Mc/s observations of Franklin and Burke (unpublished); only occurrences above a given activity are plotted, but were selected by a prior criterion to avoid biasing.

January 1957:   18-Mc/s observations of Barrow, Carr, and Smith.[12]

A slow drift in position of the most active regions can be seen over the 6-yr period, implying that the active radio regions rotate on the average somewhat more rapidly than the System II coordinate system. Before the 1957 observations of Carr *et al.* were available, Burke[13] had estimated that the period was close to $9^h 55^m 28\overset{s}{.}5$, while Gallet[14] gave $9^h 55^m 29\overset{s}{.}5$. Carr *et al.*[15] gives $9^h 55^m 28\overset{s}{.}6$, while the lines drawn in Fig. 4 have a slope corresponding to $9^h 55^m 28\overset{s}{.}8$. It is interesting to note that no rotational periods of long-term visual fea-

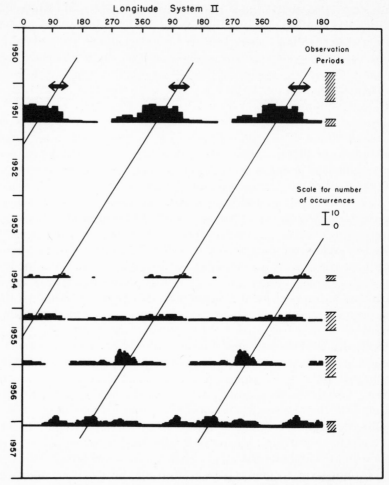

Fig. 4. Summary of 1950–51 Sydney observations, 1954–56 Washington observations, and 1957 Florida observations. The central meridian in longitude System II is shown, with lines indicating the apparent drift of a source rotating with a period of $9^h\,55^m\,28\overset{s}{.}8$.

tures have been reported in this range, indicating that the source of radio noise lies well below the visible clouds. Short periods of observation have at times suggested association with one feature or another but the long-term observations have not supported the identifications.

One further observational peculiarity of the noise bursts from the planet Jupiter can be noted from Fig. 4. It was noticed by Shain[11] that the major active region in 1951 was not visible over a full 180° of planetary rotation, as it should be if the source were an isotropic radiator. Apparently, the active region was visible only when it was within about ±70° of the central meridian. Later observations shown in Fig. 4, particularly those centered on December 1955 and January 1957, show an even narrower range of angle, the most active region being observed only within ±30° of the central meridian. It is natural to suspect that an ionosphere is responsible for this restriction, but such a mechanism would make the cone of reception strongly frequency-dependent, an effect that has not been noticed in the data available so far. This is unfortunate, for the prospect of observing the ionosphere of another planet is an exciting one, and perhaps it is not impossible that some as yet unsuspected relation might exist between the radio noise and the ionosphere of Jupiter.

The observational and interpretational problems associated with thermal and nonthermal planetary radio noise present, therefore, an interesting contrast. Interpretation of the thermal component is aided by our better understanding of the physical processes involved, but must wait for more complete, and by no means easy, radio observations. The nonthermal radiation, on the other hand, presents first of all an interpretational problem, although more observations of the complex spectral, polarization, and time-varying properties of the radiation are also required, before speculations can become theories. Observations of the thermal component require the services of the largest steerable paraboloids, coupled with the most refined receiving equipment. The noise bursts from Jupiter, on the other hand, can be observed with the simplest of antennas and receivers, although other planets, if they prove to be sources of nonthermal noise, will undoubtedly require larger antennas. In any event, the next few years are certain to provide exciting new information in both these fields, increasing our understanding of the puzzling nonthermal radio noise, and forcing upon us new interpretations of the thermal noise, where we are so confident that understanding exists.

Discussion

D. H. Menzel, *Harvard College Observatory:* The temperature of 600° from 3-cm observations of Venus is very difficult to explain in terms of equilibrium values, because it is far above anything that you would expect to be maintained by normal solar radiation. I would judge, therefore, by qualitative astrophysical reasoning, that the source must be some high region of the Venusian atmosphere that is heated by the far ultraviolet with its higher effective temperature—as in our ionosphere, where we find similar high temperatures. If such an argument could explain the high value, then the radio observations would tell us nothing about the surface or about water vapor on the surface.

B. F. Burke: I think that a very dense ionosphere is needed in order to do this by either free-free or free-bound transitions of ions and electrons. A definite model with numbers would be necessary to prove that it is possible to get appreciable optical depth with a reasonable ionosphere. A more promising approach is to ask what molecules might be present to give appreciable absorption, not necessarily in the ionosphere but in the lower levels of the atmosphere as well. I think that the answer can be given only experimentally. We will have to measure temperature as a function of frequency, and look for peaks and valleys, analogous to those in our own atmosphere at 5 mm and 1.25 cm. Concerning the 600° temperature at the surface, the greenhouse effect has been suggested, although this is also a qualitative idea. One can probably argue either way, at the present state of our knowledge.

Menzel: In any event, the radio observations do not say anything about whether or not there is water present as a substance on the planet. At most you could argue that any water is in the form of steam. Then 600° would be the temperature of the steam roughly in the region where the optical depth becomes unity.

Burke: Of course. However, if it is the surface of the planet that has this temperature there is no liquid water on the surface.

Menzel: That would be true, unless the atmosphere of Venus is many times more dense at its base than that of the earth.

O. E. H. Rydbeck, *Chalmers University, Sweden:* Are left-hand and right-hand polarization of the Jovian nonthermal signals equally frequent, or does one sign predominate?

Burke: We had only a very short series of observations, all but one of which were of one hand. It was interesting that the sense of polarization determined for the most active region was the same as that determined from a single measurement by Shain in Australia at approximately the same period (early 1956).

Note added: Seven events were observed by Franklin and Burke, of which five exhibited right-handed polarization, one left-handed polarization, while one was of mixed character, with different bursts exhibiting different senses, and some bursts polarized little if at all. The handedness refers to the radio convention, in which the sense of rotation is that observed looking in the direction of propagation.

P. E. Green, *M.I.T. Lincoln Laboratory:* Is there any correlation between the observed nonthermal activity on Jupiter and the position of the red spot?

Burke: In 1955–56 the most active region did have approximately the same longitude as the red spot, but the correlation does not appear for other years. The rotational period of the red spot is $9^h 55^m 42^s$, which is rather different from the apparent rotational period for the active radio regions derived from the analysis of several years' data.

K. Toman, *Geophysics Research Directorate, AFCRC:* Is there any sunspot-cycle effect observed in Jovian noise?

Burke: There is no evidence for a sunspot-cycle effect in the data I have given.

R. N. Bracewell, *Stanford University:* I would like to ask Dr. Burke to comment from his own observations on the possibility of nonthermal noise from Venus.

Burke: In the fall of 1956, Dr. Erickson and I operated equipment at 22 and 27 Mc/s, looking at Venus in the morning sky before the sun rose. The published information purporting to show noise from Venus has all been taken during the day when interference from the sun and from terrestrial radio stations is common, and can be criticized on this as well as other bases. By looking at Venus before sunrise, we could eliminate this source of error, and we looked for about a month, getting a good record every day, with completely negative results.

Bracewell: What was the reported recorded intensity?

Burke: The reported intensity was very large—of the order of 10^{-21} wm^{-2} (c/s)$^{-1}$, I believe. We would have detected 10^{-23} wm^{-2} (c/s)$^{-1}$.

REFERENCES

1. C. L. Mayer, T. P. McCullough, and R. M. Sloanaker, *Astrophys. J. 127*, 1 (1958).
2. F. D. Drake and H. I. Ewen, Report of URSI U. S. A. National Committee, 12th General Assembly. National Academy of Sciences–National Research Council Publ. No. 581 (1958), p. 145.
3. T. Dunham, Jr., *The atmospheres of the earth and planets*, ed., G. P. Kuiper, (University of Chicago, 1949), p. 286.
4. J. L. Pawsey and R. N. Bracewell, *Radio astronomy* (Oxford University Press, 1955), Chap. VIII.
5. A. D. Mitra and C. A. Shain, *J. Atm. Terrestrial Phys. 4*, 204 (1954).
6. B. F. Burke and K. L. Franklin, *J. Geophys. Research 60*, 213 (1955).
7. B. F. Burke and K. L. Franklin, *Radio astronomy*, ed., H. C. van de Hulst, (Cambridge University Press, 1957).
8. A. G. Little and Ruby Payne-Scott, *Australian J. Sci. Research A 4*, 489 (1951).
9. K. L. Franklin, *Carnegie Institution of Washington Year-book 55*, 75 (1956).

10. C. A. Shain, *Nature 176,* 836 (1955).

11. C. A. Shain, *Australian J. Phys. 9,* 61 (1956).

12. C. H. Barrow, T. D. Carr, and A. G. Smith, *Nature 180,* 381 (1957).

13. B. F. Burke, *Carnegie Institution of Washington Year-book 56,* 90 (1957).

14. R. Gallet, Report of URSI U. S. A. National Committee, 12th General Assembly. National Academy of Sciences–National Research Council Publ. No. 581 (1958), p. 148.

15. T. D. Carr, A. C. Smith, R. Pepple, and C. H. Barrow, *Astrophys. J. 127,* 274 (1958).

11

Correcting Noise Maps
for Beamwidth

R. N. BRACEWELL

Introduction

In the course of the first exploratory surveys of the distribution of
radio noise over the sky, the scientists concerned were acutely aware
of the inadequacy of their antennas to reveal detailed spatial structure. Such detail, of course, is a prerequisite to a physical understanding of the nature of the sources of the radiation and, even today,
to the extent that the origin of the many forms of celestial radio
emission is not understood, it is commonly lack of angular resolution
which is felt to be the limiting factor. From the beginning of radio
astronomy, therefore, attention has been paid to the question of ex-
tracting the utmost from the data obtained by scanning with an
antenna, and a good theoretical basis now exists from which to discuss the special problems of application with which this symposium
is concerned.

Two kinds of practical problems arise, which we may refer to as
"smoothing" and "sharpening." The first of these arises when we
have a detailed survey available and wish to find what we would observe with a beam wider than that used for the survey. The second
problem comes up when we are given the data from a survey made
with an antenna having a known radiation pattern and we wish to
ascertain what would have been observed had we used an antenna
with a narrower beam.

The first of these, the smoothing problem, is straightforward and
the solutions are easy to carry out. The sharpening problem, however,
is a delicate one involving a knowledge of astronomical sources as well
as intimate details of the survey instruments and procedure, and is a
matter primarily for the researcher. In the noise charts prepared at

the Harvard College Observatory the sharpening problems have been
carefully attended to during the preparation of the charts with a view
to requiring the user to apply, where necessary, only the less sensitive
of the two processes, that is, further smoothing.

In this paper we consider first a theorem which is valuable for the
handling of two-dimensional data since it enables numerical work to
be done at rather coarse discrete intervals without loss of accuracy.
Then a section is devoted to the procedure for smoothing survey data
to yield values appropriate to broader circular or asymmetric beams.
Practical numerical details are dealt with in a section on adjustment
of scale factors and finally the sharpening procedure is discussed.

The Discrete-Interval Theorem

A powerful theorem relating to noise surveys states that it is sufficient to have values on the points of a certain rectangular lattice in
order to know the noise distribution fully. This property is a two-
dimensional analog of the sampling theorem of communication theory
and is traced back ultimately to the fact that antennas are finite in
extent. Their parts are never sufficiently separated to take cognizance
of the coherence between those points in a received wave which contain the information on the Fourier components of the spatial noise
distribution whose angular period is finer than a certain limit.

Let x and y be rectangular coordinates respectively across and
along a circle of declination, and w_x and w_y the widths of the antenna
aperture after projection on a plane parallel to the xy-plane. Then an
observed noise-temperature distribution $T_a(x, y)$ is completely determined by its values at the points $(mX + a, nY + b)$, where m and n
assume all integral values, a and b are arbitrary constants, and the
spacing between the points is limited to values for which

$$X \leq \lambda/2w_x,$$
$$Y \leq \lambda/2w_y.$$

From the proof of this theorem[1] it will be found that the principal
assumptions are that the antenna may be represented by a large
plane aperture of finite extent, and that the field surveyed is contained in a narrow zone of declination not too near the poles. If the
latter assumption is not met it is necessary to break the survey down
into narrower zones, or in some other way allow for the inadequacy
of the rectangular cartesian coordinates (x, y).

The discrete-interval theorem is of great practical value. For observers it means that successive sweeps of an antenna beam need not
be more closely spaced than a limit determined by the projected

width of the antenna, and for numerical reduction it is just as important, for it permits data to be handled, without loss of accuracy, at comparatively coarse discrete intervals.

The intervals $\lambda/2w_x$ and $\lambda/2w_y$ will be referred to as the peculiar intervals.

Smoothing

Let $T(x, y)$ be a true distribution of noise temperature over the sky and let $A(x, y)$ represent the radiation pattern of a symmetric antenna with which a distribution $T_a(x, y)$ was measured. Then

$$T_a(x, y) = \int_{-\infty}^{\infty} \int_{-\infty}^{\infty} A(x\text{-}x', y\text{-}y')\, T(x', y')dx'dy',$$

an equation which, for brevity, may be written

$$T_a = A * T.$$

It is desired to find $T_b(x, y)$, the distribution which would have been observed if a broader antenna pattern $B(x, y)$ had been used, namely,

$$T_b = B * T.$$

Since the true distribution T is not fully known, only the observed T_a, the desired solution should not require knowledge of T.

Suppose that there is a function K such that

$$B = K * A.$$

Then,

$$
\begin{aligned}
T_b &= (K * A) * T \\
&= K * (A * T) \\
&= K * T_a.
\end{aligned}
$$

(We have here used the fact that convolution obeys the associative law.) Hence T_b may be obtained directly by operating on T_a in precisely the way that T_a was obtained from T, provided that, given the antenna patterns A and B, one can solve

$$B(x, y) = \int_{-\infty}^{\infty} \int_{-\infty}^{\infty} K(x\text{-}x', y\text{-}y')\, A(x', y')dx'dy'$$

for the necessary function K. Consideration of the Fourier transform of this equation shows that K is uniquely determined, save for additive null functions, provided that the transform of A is nowhere zero where the transform of B is zero. When the second antenna is made smaller than the first, the condition that a K adequate for the problem exists reduces to requiring that the smaller antenna contain no pairs of elements with vector displacements not represented in the larger. This condition is ordinarily met, exceptions occurring with antennas which are elongated and differently oriented, and with some interferometers. These exceptions will not be considered here.

As a simple example, which is immediately applicable to many practical cases, consider an antenna pattern which is Gaussian:

$$A(x, y) = \exp - \left(\frac{x^2}{2\sigma_A{}^2} + \frac{y^2}{2\tau_A{}^2} \right).$$

Although many antenna beams are nominally circular in cross section, it is desirable to allow different widths σ_A and τ_A in the x- and y-directions to allow for the asymmetry, which in fact usually does afflict nominally circular beams, and for a difference in scale factors between the x- and y-directions, which often has to be accepted. Now if

$$B(x, y) = \exp - \left(\frac{x^2}{2\sigma_B{}^2} + \frac{y^2}{2\tau_B{}^2} \right)$$

it is known that

$$B = K * A,$$

where $K(x, y)$ is also a Gaussian pattern, that is,

$$K(x, y) = M \exp - \left[\frac{x^2}{2(\sigma_B{}^2 - \sigma_A{}^2)} + \frac{y^2}{2(\tau_B{}^2 - \tau_A{}^2)} \right]$$

and

$$M = \frac{\sigma_B \tau_B}{\sigma_A \tau_A (\sigma_B{}^2 - \sigma_A{}^2)^{1/2} (\tau_B{}^2 - \tau_A{}^2)^{1/2}}.$$

This result is an immediate consequence of the fact that the two-dimensional Fourier transform of a Gaussian function is itself Gaussian, and it may be summarized by saying that, under convolution, the variances add and the integrals multiply.

It will be noted when σ_B and τ_B are hardly any wider than σ_A and τ_A, that K is a narrow, peaked function, approaching a two-dimensional impulse function in the limit. On the other hand, when B is much broader than A, K is practically the same as B.

To carry out the smoothing process numerically on actual data one calculates

$$T_b(x, y) = XY \Sigma_m \Sigma_n K(x\text{-}mX, y\text{-}nY) T_a(x, y),$$

taking the intervals X and Y in accordance with the discrete-interval theorem.

Now it is not necessary to know the antenna dimensions to choose X and Y; clearly it is sufficient to have the radiation pattern of the antenna. Even this may not be available in detail, but the principal effects of smoothing will depend mainly on the beamwidths to half-power and so if beamwidth is all that is available it is practical to match beamwidths with a Gaussian pattern, on the assumption that normal design procedures regarding the tapering of aperture distributions will have been followed. We find that the standard deviation σ of a Gaussian curve of width θ between half-peak values is given by

$$\sigma = 0.425\theta.$$

For a uniform aperture of width w, θ is equal to $0.89\lambda/w$ and the critical interval is 0.56 times the beamwidth to half-power, or

$$X = 1.32\sigma,$$

where σ is the standard deviation of the matching Gaussian curve.

While the double summation yielding $T_b(x, y)$ is much less laborious to calculate than the two-dimensional convolution integral, it is still tedious—in fact, prohibitively so if any volume of data has to be handled. A search has therefore been made for operations which would be approximate but simple to carry out. The advantage of such an approach includes that of quickly seeing the order of magnitude of the correction involved, and when the correction required is small, the approximate method described below is especially appropriate.

In one dimension one may apply, graphically or numerically, the chord construction,[2] and in two dimensions the generalization of the same procedure may be used.[3]

Let

$$T_b(x, y) = \frac{1}{2\pi\sigma\tau} \exp\left[-\left(\frac{x^2}{2\sigma^2} + \frac{y^2}{2\tau^2} \right) \right] * T_a(x, y),$$

where we have allowed for noncircular beams of all degrees of elongation by taking different standard deviations σ and τ in perpendicular directions. (If these two directions vary with respect to the x- and y-directions, interesting phenomena arise.[4])

Denoting Fourier transforms by bars,

$$\overline{T}_b(u, v) = \exp\left\{ -2\pi^2(\sigma^2 u^2 + \tau^2 v^2) \right\} \overline{T}_a(u, v).$$

Using the formula

$$\exp \theta^2 = 1 + \sin^2\theta + \frac{5}{6} \sin^4\theta + \cdots, \quad \theta^2 < \pi^2/4,$$

we find

$$\overline{T}_b(u, v) = (1 - \sin^2\pi\alpha u - \sin^2\pi\beta v + \cdots)\overline{T}_a(u, v),$$

where $\alpha = 2^{1/2}\sigma$ and $\beta = 2^{1/2}\tau$.

Taking inverse transforms, and provided that $\overline{T}(u, v)$ does not extend outside the central rectangle of breadth α^{-1} and height β^{-1},

$$T_b(x, y) = T_a(x, y) + \tfrac{1}{4}{}^\alpha\Delta_x{}^2\, T_a(x, y) + \tfrac{1}{4}{}^\beta\Delta_y{}^2\, T_a(x, y) + \cdots,$$

while the operator ${}^\alpha\Delta_x{}^2$ stands for the second finite difference taken over intervals α in the x-direction. It is thus possible, under the conditions stated, to determine T_b from T_a by differencing the data, and we propose here to use only the first two terms involving differences. The approximate solution, expressed in full, is

$$\begin{aligned}
T_b(x, y) &= T_a(x, y) + \tfrac{1}{4}[T_a(x + \alpha, y) - 2T_a(x, y) + T_a(x - \alpha, y)] \\
&\quad + \tfrac{1}{4}[T_a(x, y + \beta) - 2T_a(x, y) + T_a(x, y - \beta)] \\
&= \tfrac{1}{4}[T_a(x + \alpha, y) + T_a(x - \alpha, y) + T_a(x, y + \beta) \\
&\qquad\qquad\qquad\qquad\qquad + T_a(x, y - \beta)].
\end{aligned}$$

The operation is indeed simple, requiring only the evaluation of the mean of four values surrounding T_a at certain distances α and β in the x- and y-directions.

It is instructive to consider the preceding material from other points of view. We had found an exact connection

$$T_b = K * T_a$$

between T_b and T_a and concluded by deciding that, to some approximation, T_b was given by the mean of four values of T_a at certain points. One may now ask whether this proposed solution can be expressed in a form suitable for comparison with the convolution integral $K * T_a$, and this is clearly possible if we write the solution as

$$K' * T_a,$$

where

$$4K' = {}^2\delta(x + \alpha, y) + {}^2\delta(x - \alpha, y) + {}^2\delta(x, y + \beta) + {}^2\delta(x, y - \beta)$$

where ${}^2\delta(x, y)$ is the two-dimensional generalization of the impulse function $\delta(x)$, and has the property

$$\int\limits_{-\infty}^{\infty} \int\limits_{-\infty}^{\infty} {}^2\delta(x, y)\, dx dy = 1.$$

A convenient notation results if K' is represented by the two-dimensional array of the coefficients of the impulse functions at the intersections of the α, β-lattice. Thus

$$T_b \approx \begin{bmatrix} 0 & \tfrac{1}{4} & 0 \\ \tfrac{1}{4} & 0 & \tfrac{1}{4} \\ 0 & \tfrac{1}{4} & 0 \end{bmatrix}^* T_a.$$

Note that the sum of the coefficients of the array is unity and that the meaning of this fact is that if T_a is constant then the indicated operation will leave T_a unchanged. In terms of the two-dimensional Fourier spectrum this means that the zero-frequency component is left unaltered if the sum of the coefficients is unity.

Now that the proposed solution is in the same form as the exact solution, assessment of the accuracy obtained should be possible by comparison. At first sight it may seem that a set of four impulses is a rather drastic substitution for a single smooth Gaussian peak, but it must be remembered that T_a contains no spatial frequencies beyond certain limits. Consequently, any spatial frequencies composing the impulse tetrad which are not present in T_a will produce no effect. Therefore

$$\begin{bmatrix} 0 & \tfrac{1}{4} & 0 \\ \tfrac{1}{4} & 0 & \tfrac{1}{4} \\ 0 & \tfrac{1}{4} & 0 \end{bmatrix} * T_a = M(x,y) * \begin{bmatrix} 0 & \tfrac{1}{4} & 0 \\ \tfrac{1}{4} & 0 & \tfrac{1}{4} \\ 0 & \tfrac{1}{4} & 0 \end{bmatrix} * T_a$$

$$= \tfrac{1}{4}[M(x + \alpha, y) + M(x - \alpha, y) + M(x, y + \beta) + M(x, y - \beta)] * T_a,$$

where $M(x, y)$ is the function whose Fourier transform is constant at all the frequencies contained in T_a, and zero elsewhere. For a rectangular, uniformly excited antenna

$$M(x, y) = \frac{\sin(\pi x/\alpha)\,\sin(\pi y/\beta)}{(\pi x/\alpha)\,(\pi y/\beta)},$$

and in general it is the "principal solution" for the response of the given antenna to a point source. Hence the function with which K may be compared is the single smooth-peaked function of bounded spectral extent of which the quoted array is a set of defining samples at the peculiar intervals.

A practical method for improving the approximation is immediately suggested by this approach. If a five-point representation is used by means of the array

$$\begin{bmatrix} 0 & m & 0 \\ m & 1-4m & m \\ 0 & m & 0 \end{bmatrix},$$

a degree of freedom is available in the choice of the factor m for a best fit with K.

Consideration of the Fourier transform of K', with a view to comparison with the transform of K, leads to a very direct method of establishing or sophisticating the arrays of coefficients. In the present case the transform of K' is

$$\tfrac{1}{2} \cos 2\pi\alpha u + \tfrac{1}{2} \cos 2\pi\beta v,$$

which matches the transform of K near the origin, but, being periodic, is a poor match at high spatial frequencies. This comment, of course, is precisely the same as saying that the impulse set is spiky compared with K, but, as we have seen, this does not matter.

Adjustment for scale factors

The method given for establishing K' can be interpreted as meaning that the transforms of K and K' were matched in value and curvature at their origin. Hence, by simple properties of Fourier transforms, K and K' have the same volume and the same variances. The precise strength and spacing of the impulses composing K' are immaterial provided that the sum of the coefficients is unity and that the two variances are σ^2 and τ^2.

Suppose, as often happens, that the interval most convenient for numerical work is not precisely the interval determined by consideration of the antenna beamwidth. In a particular case an interval of $9°.6$ was indicated, but it was necessary to work at intervals of $8°$ because of previous handling of the data. To compensate the reduction in variance by a factor $(8/9.6)^2$, each coefficient 0.25 was multiplied by $(9.6/8)^2$, that is, replaced by 0.36. Now the sum of the coefficients could be kept at unity by using a central value of -0.44:

$$\begin{bmatrix} 0 & 0.25 & 0 \\ 0.25 & 0 & 0.25 \\ 0 & 0.25 & 0 \end{bmatrix} \longrightarrow \begin{bmatrix} 0 & 0.36 & 0 \\ 0.36 & -0.44 & 0.36 \\ 0 & 0.36 & 0 \end{bmatrix}$$
$$\text{(9°.6 interval)} \qquad\qquad \text{(8° interval)}$$

Only a limited adjustment can be made in this way but it is usually practical to test a proposed adjustment on the data. To examine the question theoretically, first compare the transform of the smooth function, of which the array coefficients are the samples, with the transform of K. Then take into account the extent to which the spectrum of the data extends into the region where the goodness of fit to K falls off. This can be done theoretically but the principal result of considering the matter in this way is to show that the range of adjustment for scale factors depends on the character of the data.

It often happens that unequal adjustments are wanted in the x- and y-directions—for example, as a result of changing from one declination zone to another, or because the antenna beam is noncircular. This is done by first matching variances in the two directions and then selecting the central coefficient to bring the sum to unity.

Very elongated beams are handled by the use of unequal intervals α and β which are related to the standard deviations σ and τ in the x- and y-directions by $\alpha = 2^{1/2}\sigma$ and $\beta = 2^{1/2}\tau$, as mentioned earlier.

Sharpening

The inverse operation to smoothing is a more delicate operation for the following reason. Certain information may be destroyed by smoothing and therefore cannot in general be restored; more explicitly, those Fourier components falling where the Fourier transform of the antenna pattern is zero will be missing entirely from the observed data, so that there is no information on their amplitude. In some applications—to radar, for example—it may be possible to restore lost information by incorporating extraneous knowledge of the properties of the target, but in such basic fields as radio astronomy the observed data themselves normally constitute the full knowledge available. Now in other spectral bands the information is not entirely missing but is attenuated, some of it to such levels that any attempts to restore it may introduce more noise than signal. In other words, spurious detail may be introduced.

It is therefore necessary to take into account the errors afflicting the data before a restoration procedure can be laid down for optimum results. The appropriate considerations will not be entered into but it has been shown that the method only briefly described here is conservative provided the root-mean-square errors do not exceed 15 percent of the observed level.[5] With more precision, greater care in restoration would be justified if the increased complexity were acceptable.

We shall describe a procedure here which depends on the method of successive substitution for the solution of integral equations. Let

$$T_b = K * T_a.$$

Now $K * T_b$ will be even smoother than T_b and the difference $T_b - K * T_b$ is an approximate estimate of $T_a - T_b$. Therefore, as a first approximation

$$T_a = 2T_b - K * T_b.$$

We stop at this first approximation and substitute for K the discrete array K' already discussed. Then

$$T_a = (2^2\delta(x, y) - K') * T_b,$$

or, if the smoothing array K' is

$$\begin{bmatrix} 0 & \frac{1}{4} & 0 \\ \frac{1}{4} & 0 & \frac{1}{4} \\ 0 & \frac{1}{4} & 0 \end{bmatrix},$$

then the sharpening array is

$$\begin{bmatrix} 0 & -\frac{1}{4} & 0 \\ -\frac{1}{4} & 2 & -\frac{1}{4} \\ 0 & -\frac{1}{4} & 0 \end{bmatrix}$$

and adjustments for scale factor are to be made as before.

REFERENCES

1. R. N. Bracewell, *Australian J. Phys.* 9, 297 (1956).
2. R. N. Bracewell, *J. Opt. Soc. Amer.* 45, 873 (1955); *Australian J. Phys.* 8, 200 (1955).
3. R. N. Bracewell, *Australian J. Phys.* 8, 54 (1955).
4. R. N. Bracewell, *Australian J. Phys.* 9, 198 (1956).
5. R. N. Bracewell, *Proc. I. R. E.* 46, 106 (1958).

12

A Study on
Cosmic Radio Noise Sources

DONALD H. MENZEL

1. Introduction

During the past four or five years, radio engineers have occasionally written the Harvard College Observatory requesting information concerning the level of radio noise from cosmic sources. The increasing number of such requests led me to examine the available published data and to coordinate them into a single report for the use of astronomers and engineers.

The task of assembling and evaluating the data has not been easy. Many discordances needed explanation. Some observers had introduced corrections for finite beamwidth, for point sources, or for antenna distortion. When discrepancies appeared between two sets of data, the uncorrected original observations received the higher weight in this survey. In some cases, assigning a zero point to a particular set of data constituted an additional problem. In spite of these difficulties, however, a reasonable map of radio noise levels from various parts of the sky resulted.

Then arose the problem of devising simple and practical methods by which an engineer working at a given latitude could easily derive the level of cosmic noise interference at any point in the sky. The solution involves the use of specially designed charts in the form of a planisphere, comprising a transparent sky map of levels of radio noise, which rotates over a grid representing the various altitudes and bearings in the visible sky at the observer's latitude.

2. Outline of the Program

The current study led to the preparation of maps of the distribution of sky brightness at various radio frequencies, in a form useful

151

for communications engineers. The source material consisted of published surveys of cosmic noise by various investigators. The data given in these surveys required evaluation and preliminary reduction to a common half-power beamwidth.

Analysis indicated that only two basic maps were necessary, one for 200 Mc/s and 10° beam, and the other for 600 Mc/s and 2° beam. Each map is in two parts, one covering the northern celestial hemisphere, and the other the southern celestial hemisphere. The unit chosen for the sky brightnesses on the maps, selected after consultation with many scientists and engineers, is the decibel, and the reference level is 1°K; thus

$$db = 10 \log T,$$

where T is the sky temperature for the given frequency. Comparison of published data at various frequencies provided the relation of sky brightness at the basic frequencies given above to brightness at other frequencies. Using this relation in conjunction with the basic maps, the observer may easily determine the distribution of cosmic noise at any frequency.

The radio noise maps employ equatorial coordinates on the polar stereographic projection. A series of 15 horizon-coordinate grids enables rapid determination of cosmic radio noise interference intercepted by an antenna pointing at a given altitude and bearing. These grids are designed for use by observers at latitudes 0°, ±25°, ±35°, ±40°, ±45°, ±55°, ±65°, and ±75°. The two noise maps (Fig. 2) and the 15 grids (Fig. 3) comprise a basic set.

A limited number of complete sets of grids, with 12-in. noise maps on transparent celluloid, are available at nominal cost. In place of the larger maps, one may copy those presented here and mount them as described in Sec. 5 to give the desired solution.

This study also presents a tabulation of the intensities of the discrete sources of cosmic radio radiation, as seen at various frequencies in a 2° beam. These intensities may be easily converted into intensities at any other half-power beamwidth. In addition, the analysis includes an estimate of the contributions of bodies in the solar system to the cosmic radio interference.

3. The Published Data

Table 1 summarizes the characteristics of the published measurements of the background radiation from the sky.

The sky coverage of published surveys in various frequency ranges appears in Fig. 1, which specifies the number of surveys as a function

153

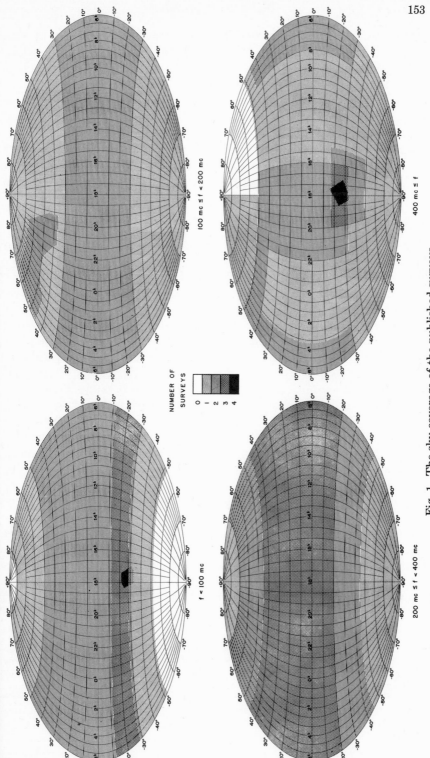

f < 100 mc

100 mc ≤ f < 200 mc

NUMBER OF
SURVEYS

0
1
2
3
4

200 mc ≤ f < 400 mc

400 mc ≤ f

Fig. 1. The sky coverage of the published surveys.

TABLE 1. Published surveys.

Ref. no.	First author	Publ. year	Freq. (Mc/s)	Wavelength (m)	Antenna beamwidth (deg)	Sky coverage[g] Right ascension (h)	Declination (deg)	Ant. effic.	Correc. for smoothing	Temp. scale	Effect of discrete sources
1	Reber	1944	160	1.9	11.6	0-24	+90--32.5	h	k	n	—
2	Sander	1946	60	5.0	35 × 30[d]	0-24	+66; +21°45'; -10	i	k	o	—
3	Hey	1946	64	4.7	30 × 12[e]	0-24	+60--30	i	l	o	—
[a]4	Hey	1948	64	4.7	14 × 13[e]	0-24	+50--30	h	k	o	—
5	Reber	1948	480	0.62	3.9	0-24	+44--32	h	k	n	—
[a]6	Bolton	1950	100	3.0	17	0-24	+30--90	h	m	o	—
7	Allen	1950	200	1.5	24	0-24	+45--90	h	m	n	—
8	Shain	1951	18.3	16	50 × 28[d]	0-24	+30--90	h	l	o	—
9	Piddington[b]	1951	1210 / 3000	0.25 / 0.10	5.8, 2.8 / 1.7	17-18.5	-20--33	h	k	n	—
10	Atanisiyevič	1952	255	1.2	9.3	l = 120°-320°	Galactic center b = +30--30	i	k	p	—
11	Brown	1953	158.5	1.9	2	l = 40°-130°	b = +14--14	h	k	o	—
[a]12	Shain	1954	18.3	16	17	0-24	-12--52	h	k	o	—
13	Higgins	1954	9.15	33	31 × 26[d]	0-24	-32	h	k	o	—
14	McGee	1954	400	0.75	2	15-20	-17--49	i	k	n	—
15	Kraus	1954	250	1.2	17 × 5 or 2.5[d]	15-21	+60--40	i	k	n	r
[a]16	Kraus	1955	250	1.2	8 × 1[d]	12-24	+65--38	j	k	n	—
[a]17	Denisse	1955	909	0.33	2.6	16-22	+53--33	i	k	q	—
[a]18	McGee	1955	400	0.75	2.2	15-20	-17--49	i	k	o	r
[a]19	Baldwin	1955	81.5	3.7	15 × 2[d]	0-24	+82--28	j	k	o	—
[a]20	Piddington	1956	600	0.50	3.3	0-24	+51--90	h	k	n	—
21	Dröge[c]	1956	200	1.5	16.8 × 16.3[e]	0-24	+90--20	h	m	o	—
22	Westerhout	1956	1390 / 400	0.22 / 0.75	2.8 × 1.9[f] / 2.1	l = 323°-353° / l = 324°-338°	b = +4--7 / b = +8--10	i	k	n / o	—
23	Shain	1957	19.7	15	1.4	l = 321°-336°	b = +5--7	i	k	o	—
24	Ko	1957	250	1.2	8 × 1[d]	0-24	+63--41	h	k	n	r
25	Denisse	1957	909	0.33	2.6	18- 6 / 4- 8	+90-+40 / +45--28	i	i	o	—

1. G. Reber, *Astrophys. J. 100*, 279 (1944).
2. K. F. Sander, *J. Inst. Elect. Eng. III 93A*, 1487 (1946).
3. J. S. Hey, J. W. Phillips, and S. J. Parsons, *Nature 157*, 296 (1946).
4. J. S. Hey, S. J. Parsons, and J. W. Phillips, *Proc. Roy. Soc. A 192*, 425 (1948).
5. G. Reber, *Proc. I.R.E. 36*, 1215 (1948).
6. J. G. Bolton and K. C. Westfold, *Australian J. Sci. Research A 3*, 19 (1950).
7. C. W. Allen and C. S. Gum, *Australian J. Sci. Research A 3*, 224 (1950).
8. C. A. Shain, *Australian J. Sci. Research A 4*, 258 (1951).
9. J. H. Piddington and H. C. Minnett, *Australian J. Sci. Research A 4*, 459 (1951).
10. I. Atanisiyević, *Compt. Rend. 235*, 130 (1952).
11. R. H. Brown and C. Hazard, *Monthly Notices Roy. Astron. Soc. 113*, 109 (1953).
12. C. A. Shain and C. S. Higgins, *Australian J. Phys. 7*, 130 (1954).
13. C. S. Higgins and C. A. Shain, *Australian J. Phys. 7*, 460 (1954).
14. R. X. McGee and J. G. Bolton, *Nature 173*, 985 (1954).
15. J. D. Kraus, H. C. Ko, and S. Matt, *Astron. J. 59*, 439 (1954).
16. J. D. Kraus and H. C. Ko, *Nature 175*, 159 (1955).
17. J. F. Denisse, E. LeRoux, and J. L. Steinberg, *Compt. Rend. 240*, 278 (1955).
18. R. X. McGee, O. B. Slee, and G. J. Stanley, *Australian J. Phys. 8*, 347 (1955).
19. J. E. Baldwin, *Monthly Notices Roy. Astron. Soc. 115*, 684 (1955).
20. J. H. Piddington and G. H. Trent, *Australian J. Phys. 9*, 481 (1956).
21. F. Dröge and W. Priester, *Z. Astrophys. 40*, 236 (1956).
22. G. Westerhout, *Bull. Astron. Netherlands 472*, 105 (1956).
23. C. A. Shain, *Australian J. Phys. 10*, 195 (1957).
24. K. C. Ko and J. D. Kraus, *Sky and Telescope 16*, 160 (1957).
25. J. F. Denisse, J. Lequeux, and E. LeRoux, *Compt. Rend. 244*, 3030 (1957).

a Reference used to construct isophotal maps or to determine spectral law in this study.

b According to Ref. 20, the 1210-Mc/s results in this paper are not trustworthy.

c The fact that the isotophotes show a large number of "fingers" running east and west suggests the presence of calibration errors dependent on declination.

d Long axis directed along the meridian.

e Long axis oriented at right angles to the meridian.

f For galactic longitudes less than 3.29°.5, the major axis of the beam was perpendicular to the galactic equator; for longitudes greater than 328°.5, the major axis was parallel to the galactic equator.

g The region covered by the survey is within the stated limits. Except where otherwise indicated (l, b), the limits are in equatorial coordinates.

h One or more causes of antenna inefficiency are taken into account (copper loss, ground loss, side radiation other than that accompanying the principal beam).

i Causes of antenna inefficiency not mentioned.

j Causes of antenna inefficiency are mentioned but not taken into account.

k No restoration of data.

l Only restored data given.

m Both restored and unrestored data are given.

n Relative to the coldest observed region of the sky.

o Absolute.

p Units not given for the chart contours in this survey.

q Absolute. Reference 25 gives the value of 1 unit on the chart as 2°.5K.

r The effect of discrete sources has been corrected for.

of right ascension and declination for the following frequency intervals: $f < 100$ Mc/s; 100 Mc/s $\leq f < 200$ Mc/s; 200 Mc/s $\leq f < 400$ Mc/s; 400 Mc/s $\leq f$.

4. Reduction of the Data

The Observed Temperature Profiles

Although some of the papers listed in the previous section express the data in units of absolute brightness or temperature, others, as indicated, merely give intensities above the coldest part of the sky. Reduction of the data to a common zero, therefore, requires a knowledge of the brightness of the coldest parts of the sky as a function of frequency. Piddington and Trent[1] have derived the radio spectrum of the coldest parts of the sky ($\alpha = 09^h50^m$, $\delta = +22°$; $\alpha = 04^h30^m$, $\delta = -37°$) from observations by several people.

The published intensities were reduced to brightness temperatures, T_b, by the Rayleigh-Jeans law:

$$T_b = B\lambda^2/2k,$$

where $B\,[\text{wm}^{-2}(\text{c/s})^{-1}\,\text{sterad}^{-1}]$ is the measured brightness in the wavelength, and k is the Boltzmann constant, in this case 1.38×10^{-23} js deg^{-1}. Where necessary, the appropriate temperature of the coldest part of the sky, from the derivations of Piddington and Trent, was added to reduce the data from a given paper to a common zero.

The published isophotal charts of the references actually used here gave the basic data. From these charts we read off right ascension, along a circle of constant declination, corresponding to each brightness contour. Wherever possible, in the interest of accuracy, these chosen declination circles coincided with those along which the authors originally made their observations. When the authors failed to indicate the declinations originally used, the values selected for this survey lay at intervals of about one half of the half-power beamwidth of the original survey. Brightness temperatures of the contours, plotted as a function of right ascension for each declination, yielded a series of observed temperature profiles.

Correction for Antenna Pattern

The observed profiles required correction for antenna beamwidth to give the corrected temperature profiles. "Corrected profile" means the profile that an antenna of standard beamwidth would have given. The correction requires a "smoothing" process when the beamwidth used in the published survey is smaller than the standard beamwidth; or a "sharpening" when the original beamwidth is larger.

Bracewell and Roberts[2] and Bracewell[3] have discussed in detail methods of correction for antenna smoothing. The method used here, outlined by Bracewell, involves replacing the true antenna pattern by a Gaussian pattern of the same half-power beamwidth, and calculating, for the observed brightness temperature at a point, a correction based on temperature at four points surrounding this point.

Let T_c denote the observed temperature at (α, δ). Read the values T_{a1} and T_{a2}, the temperatures at $(\alpha, \delta + I)$ and $(\alpha, \delta - I)$, where I is the interval in declination. Also note T_{b1} and T_{b2}, the temperatures at $(\alpha + I \sec \delta, \delta)$ and $(\alpha - I \sec \delta, \delta)$, where I is the adopted interval in right ascension at the celestial equator. Let T_o, the corrected temperature at (α, δ), be

$$T_o = cT_c + b(T_{b1} + T_{b2}) + a(T_{a1} + T_{a2}).$$

The constants, $a, b,$ and c are determined as follows. Let B denote the half-power beamwidth in right ascension of the antenna used in the published references; B , the half-power beamwidth of this antenna in declination; and B_s, the standard half-power beamwidth, equal in right ascension and declination, to which we desire to reduce the published data. Determine

$$\sigma'_\alpha = (B_\alpha{}^2 - B_s{}^2)^{1/2}/2.36,$$
$$\sigma'_\delta = (B_\delta{}^2 - B_s{}^2)^{1/2}/2.36.$$

Then the constants $a, b,$ and c are

$$a = -(\sigma'_\delta / I_\delta)^2/2,$$
$$b = -(\sigma'_\alpha / I_\alpha)^2/2,$$
$$c = 1 - 2(a + b).$$

When the data from a published reference are being sharpened, $a < 0$, $b < 0$, and $c > 0$. If they are being smoothed, $a > 0$ and $b > 0$.

As an example, consider the survey made by Dröge and Priester (Ref. 21) at 200 Mc/s. They used an antenna with a 16° beamwidth. We wish to reduce the survey to a 10° beamwidth. Thus,

$$\sigma'_\alpha = \sigma'_\delta = (16^2 - 10^2)^{1/2}/2.36 = 5°.3.$$

In practice, one should choose I_α equal to about $1.5\sigma'_\alpha$, and I_δ about $1.5\sigma'_\delta$. In this case, $I_\delta = 8°$ and $I_\alpha = 30^m = 7°.5$ at the equator. Then

$$a = -(5.3/8.0)^2/2 = -0.22,$$
$$b = -(5.3/7.5)^2/2 = -0.25,$$
$$c = 1 - 2(-0.22 - 0.25) = +1.94,$$

and

$$T_o = 1.94T_c - 0.25(T_{b1} + T_{b2}) - 0.22(T_{a1} + T_{a2}).$$

For each published reference used, one derives T_o as a function of α for the various circles of declination.

The Isophotal Maps

As already mentioned in Sec. 2, the units of the isophotal contours on the final maps are decibels above $1°K$. The temperature was derived from the desired brightness level by the equation $db = 10 \log T$. The right ascension corresponding to this temperature was then read off from each corrected temperature profile. On the stereographic charts, a point was placed at the appropriate right ascension and declination, and the points of constant db were connected to form the isophotal contours.

This paper presents two standard isophotal maps of the sky: one with a frequency of 200 Mc/s and standard beamwidth 10°, and the second with a frequency of 600 Mc/s and standard beamwidth 2°. The 200-Mc/s map was derived from Refs. 7 and 21. The former covers mainly the southern hemisphere of the sky, and the latter the northern hemisphere. Since the regions covered by the two references overlap by about 60°, fairly extensive intercomparisons are possible. The data in Ref. 7 are given in intensity units above the coldest part of the sky; a zero-point correction of 77 K deg[1] was therefore added to the derived temperatures. Comparison of the data from Ref. 7 with those from Ref. 21, for which no zero-point correction is necessary, indicated that a discrepancy of 70 deg still existed, in the sense that the values from Ref. 7 were too low. This conclusion agrees well with the differences between the two references as derived by Dröge and Priester themselves (Ref. 21). Thus, the total zero-point correction in the data from Ref. 7 amounted to 147 deg. With this correction applied, the two surveys then agreed to within 1 db in the overlap region, and a mean of the two was taken in this region.

For the 200-Mc/s, 10°-beamwidth map, an attempt to remove the discrete radio sources from the data seems scarcely worth while. The two most intense discrete sources, Cassiopeia A and Cygnus A, are the only ones that contribute a significant fraction of the observed intensities at their positions. Since their spectra correspond closely to that of the background, the error introduced by not removing these discrete sources will be small.

The 600-Mc/s, 2°-beamwidth map was derived from Ref. 20. A zero-point correction of 5.7 deg was added to the temperatures given in this reference. The isophotal contours for declinations north of $+51°$ are taken from Ref. 25 (909 Mc/s). For the 600-Mc/s map six discrete sources were removed from the data: Cassiopeia A, Cygnus A, Taurus A, Virgo A, Centaurus A, and IC 443. Lettered points show the positions of these sources on the map.

Figure 2 gives the final isophotal maps of the sky, in equatorial coordinates. Each map consists of two parts, one covering the northern hemisphere of the sky, south to $-20°$ declination, and the second covering the southern hemisphere north to $+20°$ declination. On the 200-Mc/s, 10°-beamwidth map, the contour of least intensity is 23 db, and the contour interval 1 db. On the 600-Mc/s, 2°-beamwidth map, the contour of least intensity is 10 db, and the contour interval 3 db. Table 2 gives positions and identifications of the discrete sources lettered on the 600-Mc/s map.

TABLE 2. Identifications and positions of the discrete sources.

Designation	Source	Right ascension	Declination
A	Cassiopeia A	23^h21^m	$+58°.5$
B	Cygnus A	19 58	$+40.6$
C	Centaurus A	13 22	-42.8
D	Taurus A	05 32	$+22.0$
E	Virgo A	12 28	$+12.7$
F	IC 443	06 14	$+22.6$

Each map shows the ecliptic, with the position of the sun indicated at half-month intervals. The sun's contribution to the radio noise at the frequencies under discussion will be mentioned in Sec. 5.

The polar stereographic projection used in these charts is the projection of a sphere upon a tangent plane from a point on the sphere diametrically opposite to the point of tangency. In common with other azimuthal projections, the stereographic possesses the following properties:

1. All great circles passing through the center of projection appear as straight lines on the map and indicate their correct bearings, or azimuths; hence the name. (On the earth, for polar stereographic projection, lines of longitude are straight lines radiating from the center at the correct bearings. On the sky map of the Cosmic Noise Survey, the radial lines are hour angles (15° to 1 hour) which correspond, on the celestial sphere, to longitude on earth.)

2. On all polar azimuthal projections, circles of constant latitude on the earth (declination on the celestial sphere) appear as concentric circles centered on the center of projection.

3. All areas equally distant from the center have similar distortion. The stereographic projection has additional properties which are particularly useful in these noise charts:

4. All circles, regardless of size, remain circles on the map, and thus all parallels and meridians appear as arcs (meridian and hour circles degenerating to radial straight lines in the polar case).

5. All meridians lie at right angles to the parallels and their scale relation is correct in any small area.

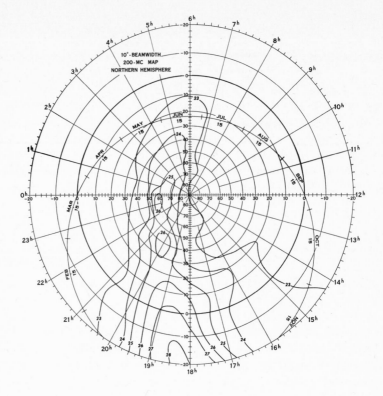

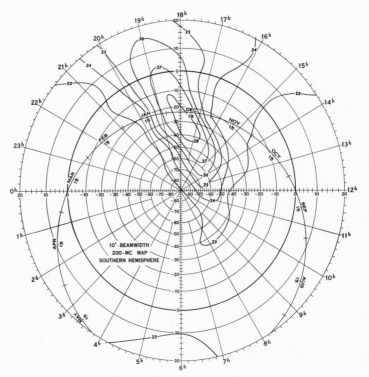

Fig. 2. The radio sky maps.

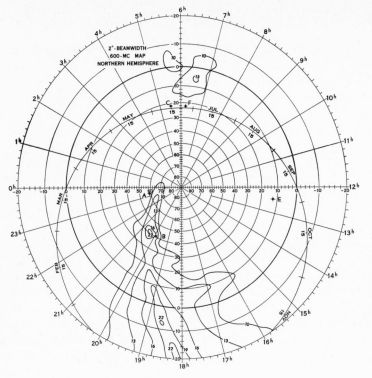

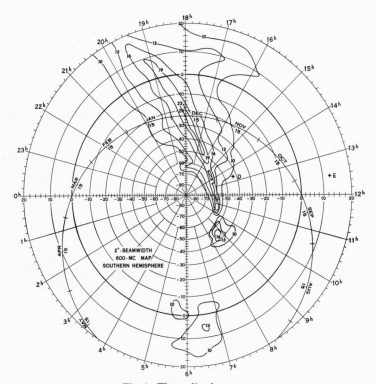

Fig. 2. The radio sky maps.
(Continued)

These last special properties indicate that the projection is "conformal." It follows that all angles are correctly represented, and that every *small* triangle is represented by a similar triangle. In fact, *all small* figures are represented by similar figures. This property is particularly relevant to the mapping of antenna patterns on the noise chart.

If the user has a circular antenna pattern with a half-power beamwidth of 2° or 10°, he need only read the noise signal in decibels above 1°K directly from the appropriate *point* on the chart, and convert it to his frequency by the nomogram provided. All of the scaling of circles for different declinations has been done for him in the construction of the map. Also, if he has a nearly circular pattern with beamwidths not too different from 2° and 10°, he can, by careful study of both sets of noise charts and some experimental work, choose a factor which should satisfactorily represent his antenna. However, many antenna patterns depart considerably from 2° and 10° circles, especially at the lower frequencies.

Bracewell, elsewhere in this volume, has developed a numerical method of "blunting" the chart beamwidths in terms of "critical intervals" equal to approximately one half of the half-power beamwidth along orthogonal axes. Whatever method one employs, the user of these charts must determine for himself how the patterns are transformed when projected on different regions of the chart.

If an antenna pattern remains fixed relative to the earth, its center sweeps out circles of constant declination on the celestial sphere and consequently the pattern on the chart need be drawn only once, for it does not change if the declination remains constant. Clearly, then, if the antenna pattern moves with respect to the earth, it must be drawn on the sky chart differently for each declination.

Circular patterns are the easiest to draw, since they remain circles everywhere on the chart, only changing scale with declination. The proper scale can be found in degrees of declination and hours of right ascension (1 hour = 15°) on the transparent charts. Although the circle is drawn as a circle, its center cannot be located where the center of the beam actually falls on the celestial sphere because the interior area of the circle is distorted. Hence, the correct procedure is to plot the end points of any diameter (vertical or horizontal generally being simplest), find the center on this diameter, and then draw the circle. Ellipses can be drawn with fair accuracy with the aid of a similar procedure for the major and minor axes. Since almost any antenna pattern can be represented reasonably well by a combination of circles and ellipses, the foregoing method covers most cases likely to occur in practice.

The Determination of the Spectral Law

The maps presented in Fig. 2 give, for each of two standard beam-widths, the distribution of intensity over the sky at a single frequency. To enable the user to convert from intensity at these frequencies to intensity at any other frequency, we need to know the spectral law, that is, the variation of intensity with frequency. The spectral law followed from a matching of isophotal contours derived from data at different frequencies to the isophotal contours on the standard maps. Specifically, for 200 Mc/s, Refs. 4, 6, 12, 16, 17, 19, 20 and 25, reduced to a beamwidth of 10°, yielded the significant data. Similarly, for 600 Mc/s, the data came from Refs. 17, 18, and 25, reduced to a beam-width of 2°. The comparison of contours at different frequencies gave the interesting result that, for a given standard beamwidth, the contours at various frequencies were the same shape, to within the accuracy of the present study.

The matching of the various data to the 10°-beamwidth map yielded a relation between the intensity at a given frequency and the intensity at 200 Mc/s. The analysis led to the spectral law:

$$db_f = db_{200} + k \log(200/f),$$

where $k = 26, f < 200$ Mc/s, $k = 2.7(35 - db_{200}), f > 200$ Mc/s.

No definitive law is derived for the 2°-beamwidth map, because of the paucity of data. However, the available surveys appear to fit the law given for the 10°-beamwidth maps.

The sharp break in the spectral law at 200 Mc/s is undoubtedly not real, but an effect introduced by the methods of analysis. However, for the purposes of this work, the law derived above appears to be sufficiently accurate.

A nomogram allowing rapid conversion from intensity at any frequency to intensity at any other frequency appears in Sec. 5.

5. The Use of the Radio Noise Maps

The Horizon-Coordinate Grids

Figure 3 presents a series of horizon-coordinate grids for use with the radio noise maps. To use the maps, proceed as follows. From an ordinary geographic map or some other source (your local post office, for example) find the geographic latitude of your antenna, to the nearest degree. Each of the 15 grids in Fig. 3 is labeled with a specific latitude. From these grids, select and copy the two whose labeled latitude corresponds most nearly to your own. (Only one grid is neces-

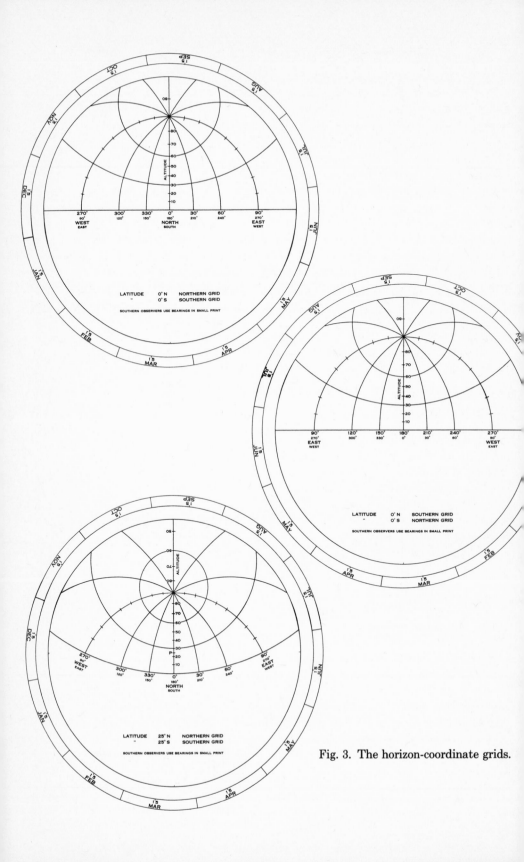

Fig. 3. The horizon-coordinate grids.

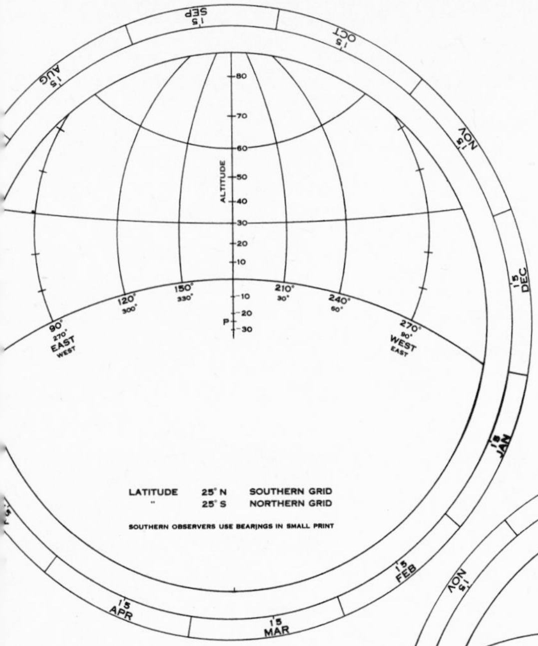

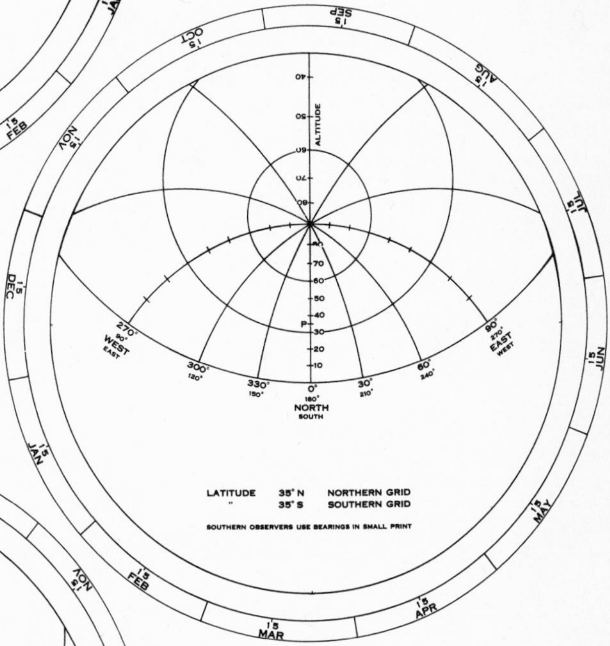

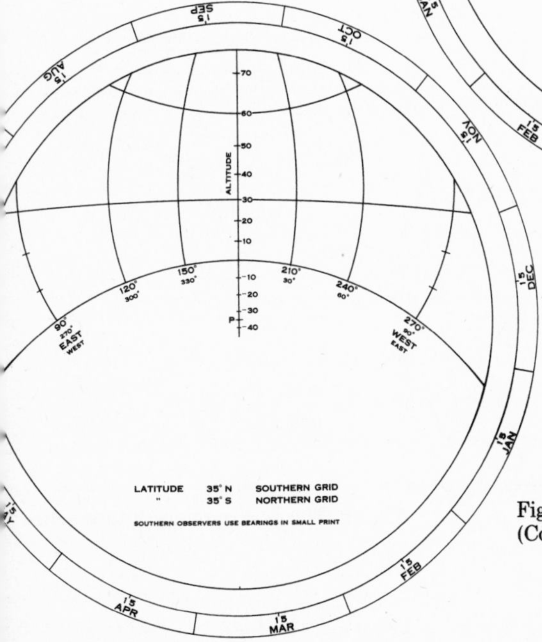

Fig. 3. The horizon-coordinate grids.
(Continued)

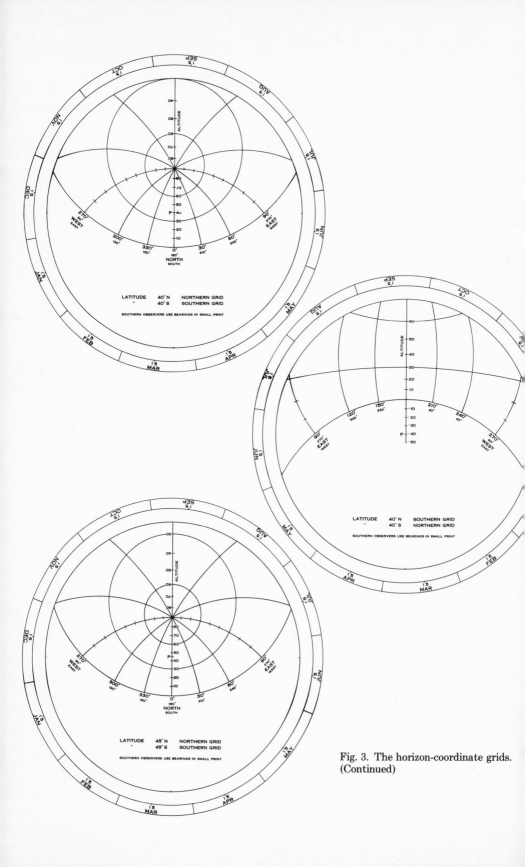

Fig. 3. The horizon-coordinate grids.
(Continued)

167

Fig. 3. The horizon-coordinate grids.
(Continued)

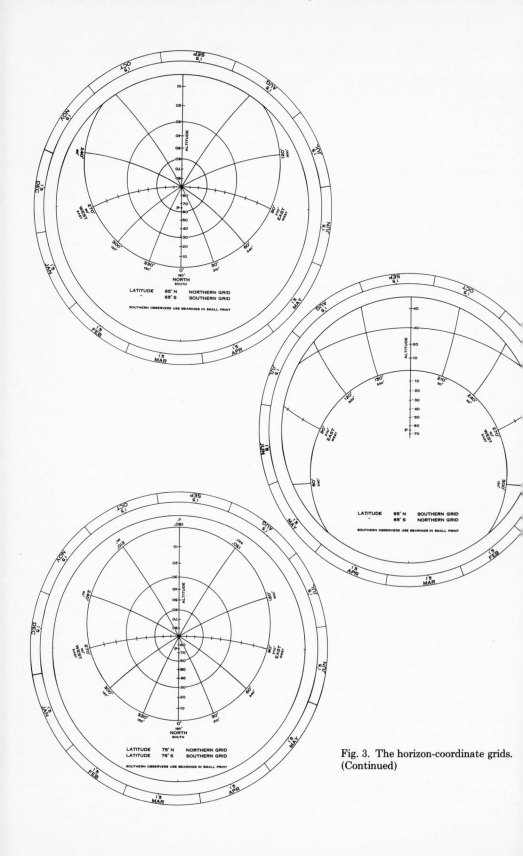

Fig. 3. The horizon-coordinate grids.
(Continued)

sary near latitude 75°.) At the center of each grid, find an index scale
marked at a given point with the letter P. Note that this letter stands
against the latitude for which the chart is labeled. Now, on the same
scale, find your own latitude. If it differs from the labeled value indicated by the letter P, mark in pencil the point on the scale that corresponds to your latitude. Carefully punch a hole in the grid through
this point.

Now choose the pair of noise maps whose half-power beamwidth,
2° or 10°, corresponds most closely to that of your own antenna. Copy
these maps on transparent cloth or paper. Place each map on its appropriate grid, according to the directions printed on the grid. Align
the center of the map with the hole you have punched in the grid.
Insert a pivot through the hole and the center of the map to fasten
the map and grid together.

The grids show the horizon, curving from east to west. The figures
along the horizon denote the bearings, in the usual sense, measured
eastward from the north. Curves of constant bearing extend from
these points. The scale along the central line indicates the altitude
above the horizon, with curves delineating the altitudes 0° (the horizon), 30°, and 60°.

Each pair of grids applies to south as well as to north latitudes.
Southern observers, however, should employ the bearings that are
printed in small type.

You are now ready to determine the diurnal variation in noise for
a given antenna. On the grid put a pencil mark to designate the bearing and altitude of the antenna beam. From the outer circle of the
grid, extend a short pencil line radially toward the center, at the position of the given day.

The time scale on the circumference of the noise maps refers to
local solar time, when the maps are used in the following manner.
Rotate the map until the local solar time on the hour scale coincides
with the penciled date line of the grid. From the curves on the noise
map, read off the value of the cosmic noise, at the position of the
antenna. Rotation of the map is equivalent to the apparent rotation
of the sky. To determine the diurnal variation of noise, make measurements covering the entire 24 hours.

Nomograms for the Conversion of Coordinates

Figures 4(a) and 4(b) present nomograms for determining declina-
tion δ and hour angle h, given altitude A, bearing β, and latitude ϕ.
The procedure for use follows. On Fig. 4(a), place a straightedge
through the intersection of the lines of desired altitude and latitude,
and the desired bearing. Read off the declination at the intersection

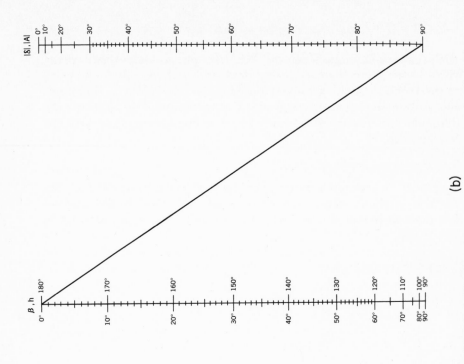

(b)

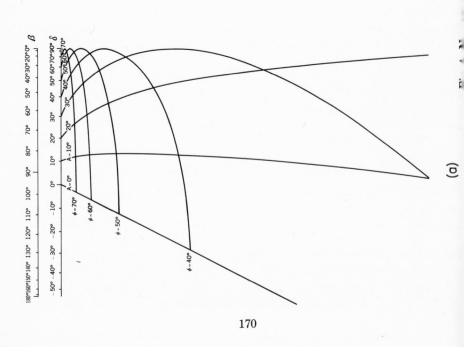

(a)

170

with the straightedge. Then, on Fig. 4(b), put a straightedge on the given bearing and the declination determined above. Mark the point of intersection of the straightedge and the diagonal line. Then put the straightedge through the given altitude and the point on the diagonal line. Read off the hour angle at the intersection with the straightedge.

The Spectral-Law Nomogram

Figure 5 presents a nomogram for converting intensity at 200 Mc/s to intensity at any other frequency, according to the spectral law derived in Sec. 4. One pair of noise maps refers to radiation at

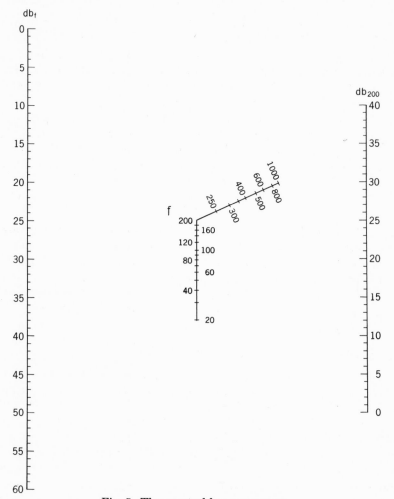

Fig. 5. The spectral-law nomogram.

200 Mc/s. Denote the noise read from these maps as db_{200}. To find the noise, db_f, for any other frequency, employ Fig. 5 as follows: Place a straightedge through the measured db_{200} and the new frequency f (Mc/s), and read off the db_f at the intersection with the straightedge.

The other pair of maps refers to noise at 600 Mc/s. Now, to convert the intensity scale to some other frequency, use the same nomogram in two steps. First, place the straightedge through the db_{600} on the left-hand scale and also through 600 on the f-scale. Then read off db_{200}, on the right-hand scale. Finally, use this value of db_{200}, as previously described, to find db_f.

The Discrete Sources

Table 2 has presented the identifications and positions of the more important discrete sources of radio noise, lettered on the 600-Mc/s map. Table 3 gives the intensity, in decibels, of these sources as seen in a 2° beam, as a function of frequency.[4] To convert the decibels in Table 3 to decibels at any other half-power beamwidth, B_o, add to the tabulated intensity a quantity $6.0 - 20 \log B_o$.

Discrete source B lies within the isophotal contours on the 600-Mc/s map. Combining the intensity of a discrete source with that from the general background requires a special procedure. Let $db_>$ be the greater of the two intensities that the observer wishes to combine, and $db_<$ be the lesser. Also, let

$$db_T = db_> + \Delta db,$$

where db_T is the combined intensity of $db_>$ and $db_<$. Table 4 presents the quantity Δdb in terms of $(db_> - db_<)$. To determine a combined intensity, find the difference $(db_> - db_<)$, enter Table 4 with this figure; read off Δdb, and add this quantity to $db_>$ to get db_T.

For example, suppose that $db_> = 10.5$ and $db_< = 8.0$. Then, from Table 4, $\Delta db = 1.9$ and $db_T = 12.4$.

Contributions by Bodies in the Solar System

The sun. The sky maps indicate the sun's position at different dates in the year as it moves along the ecliptic. The sun is an intense, variable radio source, and observers must consider its effects. Although exact statistics are not yet available, one may obtain a rough estimate of the amount of radio noise to be expected from the sun.

In the frequency range considered in this study, the contribution from the sun can be divided into three levels of activity. Figure 6 shows the approximate intensity ranges corresponding to these levels as a function of frequency as seen in a 2° beam. The lowest level, the

TABLE 3. Intensities (db) of the discrete sources.

log f	f(Mc/s)	A	B	C	D	E	F
1.30	20.0	65.5	62.4	—	57.2	60.1	—
0.35	22.4	64.7	62.0	—	55.6	58.1	—
.40	25.1	63.8	61.2	—	54.1	56.2	—
.45	28.2	62.8	60.3	—	52.7	54.3	—
1.50	31.6	61.7	59.4	—	51.3	52.5	—
1.55	35.5	60.4	58.3	—	49.9	50.7	—
0.60	39.8	59.0	57.0	—	48.4	49.0	—
.65	44.7	57.6	55.7	—	46.9	47.3	—
.70	50.1	56.1	54.3	—	45.4	45.6	—
1.75	56.2	54.7	53.0	—	44.0	43.9	37.1
1.80	63.1	53.3	51.6	—	42.7	42.3	36.0
0.85	70.8	51.9	50.2	—	41.4	40.7	34.8
.90	79.4	50.5	48.8	—	40.1	39.2	33.6
1.95	89.1	49.0	47.3	—	38.8	37.8	32.4
2.00	100	47.5	45.8	42.0	37.5	36.4	31.3
2.05	112	46.1	44.3	40.6	36.2	35.0	30.1
0.10	126	44.7	42.9	39.2	34.9	33.6	29.0
.15	141	43.2	41.5	37.8	33.7	32.2	27.8
.20	158	41.7	40.1	36.4	32.5	30.8	26.6
2.25	178	40.2	38.7	35.0	31.3	29.4	25.5
2.30	200	38.8	37.4	33.7	30.1	28.0	24.3
0.35	224	37.4	36.0	32.3	28.9	26.6	23.1
.40	251	36.1	34.6	31.0	27.7	25.2	21.9
.45	282	34.8	33.2	29.6	26.5	23.8	20.7
2.50	316	33.4	31.8	28.2	25.3	22.5	19.5
2.55	355	32.1	30.4	26.9	24.1	21.2	18.3
0.60	398	30.7	29.0	25.5	23.0	19.9	17.1
.65	447	29.3	27.5	24.1	21.9	18.6	15.9
.70	501	28.0	26.1	22.8	20.8	17.3	14.7
2.75	562	26.7	24.7	21.4	19.7	16.0	13.5
2.80	631	25.3	23.3	20.0	18.6	14.7	12.3
0.85	708	24.0	21.9	18.7	17.5	13.4	11.1
.90	794	22.6	20.5	17.3	16.5	12.1	9.9
2.95	891	21.3	19.2	16.0	15.4	10.9	8.8
3.00	1000	20.0	17.8	14.7	14.4	9.8	7.7

radiation from the "quiet sun," gives the background level of radio noise from the sun. The intensity of this radiation may decrease by a factor of 2 (3 db) from times of maximum activity on the sun to times of minimum activity.

During times of minimum activity on the sun, levels of radio-frequency radiation above that of the quiet sun have a probability of occurrence that is virtually zero. However, during periods when the sun is very active, higher levels of radio radiation must be considered.

DONALD H. MENZEL

TABLE 4. Addition of intensities expressed in decibels.

$db_> - db_<$	Δdb	$db_> - db_<$	Δdb
0.0	3.0	6.0	1.0
0.5	2.8	6.5	0.9
1.0	2.5	7.0	0.8
1.5	2.3	7.5	0.7
2.0	2.1	8.0	0.6
2.5	1.9	9.0	0.5
3.0	1.8	10.0	0.4
3.5	1.6	11.0	0.3
4.0	1.4	12.0	0.3
4.5	1.3	13.0	0.2
5.0	1.2	16.0	0.1
5.5	1.1		

Noise storms from the sun at radio frequencies consist of either a long series of separate bursts of radiation, or a general enhancement of the level of continuous radiation. Noise storms may continue for hours or days. Figure 6 shows the range of intensities of radiation from the sun at the time these storms occur.

A much shorter-lived phenomenon, the outburst, lasting for periods of the order of minutes, may raise the level of solar radio radiation considerably higher. Figure 6 indicates the general level of intensities to be expected during outbursts. The upper limit shown here may be exceeded by an occasional, very intense outburst.

The moon. As seen in a 2° beam, the moon has an intensity of about 12 db at all frequencies. To convert to other beamwidths, use the procedure described above. The moon will always lie within 5° of the ecliptic.

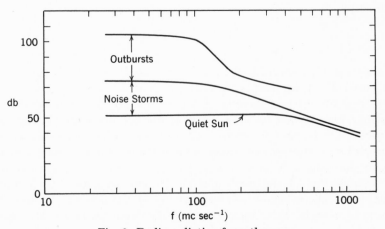

Fig. 6. Radio radiation from the sun.

The planets. The planets will always lie within a few degrees of the ecliptic. Jupiter is an intense, variable source of radio-frequency radiation. Measurements by Kraus[5] at 27 Mc/s indicate that bursts of radiation may reach an intensity of 71 db, or greater, as seen in a 2° beam. The intensity decreases very rapidly with increasing frequency. According to measurements by Smith,[6] the intensity is less than 33 db at 38 Mc/s and less than 11 db at 81.5 Mc/s, both as seen in a 2° beam. Venus, Saturn, and Mars also emit radiation at radio frequencies. However, this radiation is too weak to provide serious interference with communications.

Sources of Error

The observer must recognize that several factors, too variable for consideration here, such as ionospheric refraction and absorption, and ground reflections, also influence the intensity of radio noise interference.

6. Further Needs

The approximate nature of much of the material used in this analysis limits the accuracy of the results to the order of ±3 db. Much of the uncertainty lies in the determination of the zero point; the actual contour profiles are less uncertain. Furthermore, differences may appear between observations made in the Northern and in the Southern Hemispheres. At present, these differences cannot be resolved because the survey areas do not overlap to any great extent. Noise observations at various frequencies, made from a site near the equator, are necessary to provide satisfactory calibration between the hemispheres.

———————

Dr. Martha L. Hazen acted as chief assistant in this project, replacing Dr. Campbell M. Wade who initiated the work before he left for Australia. I am deeply indebted to both for their interest in the work and for the painstaking care they used in determining the corrections for antenna pattern. I gratefully acknowledge the help and advice of Dr. Ronald Bracewell, of Stanford University, who acted as consultant to the project, especially in the problem of data smoothing or sharpening. Miss Margaret Olmsted devoted considerable time to organization of the data and to the calculations. The following computers also worked on the project: Miss Barbara Bigelow, Mr. Courtney Doyle, Mrs. May Kassim, Mr. Richard Levy, Mrs. Mary Jane Wade, and Mr. Andrew Young. I also wish to thank Dr. H. C.

Willimantic State College Library
WILLIMANTIC, CONN

Ko, of Ohio State University, for furnishing material in advance of publication.

This research was supported in part by the Department of Defense under Contract No. DA49-170-sc-2386.

REFERENCES

1. J. H. Piddington and G. H. Trent, *Australian J. Phys. 9,* 488 (1956).
2. R. N. Bracewell and J. A. Roberts, *Australian J. Phys. 7,* 615 (1954).
3. R. N. Bracewell, *Australian J. Phys. 8,* 54 (1955).
4. J. D. Kraus and H. C. Ko, *Celestial radio radiation,* Ohio State University Research Foundation (Columbus), 1957, pp. 64–69.
5. J. D. Kraus, *Proc. I. R. E. 46,* 266 (1958).
6. F. G. Smith, *Observatory 75,* 252 (1955).

13

Interstellar Hydrogen

THOMAS GOLD

Radiation at the frequency of 1420 Mc/s is well known by now as the radiation associated with the hyperfine splitting of the ground state of the hydrogen atom. It is the frequency corresponding to the precession of the spinning electron in the magnetic field of the nucleus of the atom. There is a great deal of neutral hydrogen in the galaxy; a few percent of the entire mass of the galaxy is in this form. But, even so, the radiation at this frequency is very weak. In terms of antenna temperature, the range that can be investigated goes from 1° or 2° to something below 100°. The power is so low that this type of radiation is not likely to be a significant source of disturbance to the radio engineer; it certainly is not at the present time. With masers and parametric amplifiers the situation may change, but this type of radiation will still be very small compared with other radio emissions from the sky.

For astronomical researches this type of radiation has been of the utmost value. It so happens that most other gases can be more easily detected by spectroscopic means, but hydrogen, by far the major constituent, largely escaped detection. Tenuous neutral hydrogen gas is practically undetectable spectroscopically. It is now possible, by means of 21-cm radiation, to find the distribution in space and the state of motion of this gas. In our galaxy we have been able to plot the distribution of gases as a function of radial velocity; with this and the plausible model of the rotation of our galaxy, we have been able to draw a diagram of the gas galaxy, as it were, as it would look if we stood outside the galaxy and looked down upon it. It has also been possible to investigate particular regions in the galaxy in which interesting events are taking place. Perhaps the most exciting line of progress at present concerns regions where we are sure that within the recent past new stars have been formed. In such regions we see a lot of stars which we certainly know to be only a few million years old, very young on the cosmic time scale, and we can now investigate the associated gas. This gas may have been left over in the process of star formation, or it may be just getting ready to form more stars in the region, if indeed the process takes a little while. By understanding the distribution and the motion of the gas and of the new stars in such places, we may be able to understand how star formation occurs.

Investigations at 21 cm are not confined to our own galaxy. We can look at a few nearby galaxies also, as yet not very well, but well enough to measure the amount of gas they contain and also their approximate state of rotation. It is a particular convenience of this wavelength of the line emission as compared with other forms of radio astronomy that as soon as one can observe it, one can deduce the quantity of material in the beam (provided it is not optically thick, and usually it is not), and from a knowledge of the frequency determine the radial velocity of the material.

For observations of the universe on the largest scale, this would also be the most convenient type of information to have. In order to plot out the large-scale structure and motion, what could be better than to observe the most basic and widespread type of material, and immediately deduce the quantity of it in each interval of radial velocity. It will be the simplest kind of large-scale survey, requiring a far shorter line of reasoning for its interpretation than the optical case, where the quantity of material is not directly given by the luminosity. But how can all this be achieved?

The new and very much more sensitive receiving devices, the masers and the parametric amplifiers, together with the new large radio antennas that are now being built, will help us to reach out at 21 cm as far as optical telescopes do at present. Thus with the simple and basic type of information they provide, these techniques hold out great promise for deciphering the great puzzle of cosmology.

It would be difficult to overemphasize the extreme importance that this wavelength has to astronomy and perhaps to all the natural sciences. I wish to stress what a great pity it would be if, for reasons of carelessness or lack of information, various needless man-made disturbances were allowed to spoil the wavebands in this vicinity for the sensitive receivers that will shortly be used. I hope that this consideration will always be in mind when the frequency of any new transmitting device is chosen, and also when a decision is made on the waveform or freedom from harmonics of lower-frequency transmitters that may come into this band. At the present time radio interference is already quite severe and it is growing every year; much of it is already too close to the 21-cm band. With the more sensitive receivers we will even now have a great deal of trouble in obtaining the full sensitivity. Despite the narrow beams that are used, the almost complete absence of ionospheric reflection, and the sharp cut-off provided by the horizon, even at 21 cm it is already difficult to find good sites from which a maser could be used to full advantage.

I am reminded in this connection of the situation that has arisen in the region of frequencies around 20 Mc/s, for example. Here, in the neighborhood of the lowest frequencies that can be transmitted through the ionosphere, radio astronomers would be very eager to make many detailed observations, but they are frequently prevented by the general level of interference. When the day comes that one can easily mount substantial amounts of radio equipment in satellites that operate well above the ionosphere, it is quite likely that we will decide to use for our astronomical observations there wave bands that are too much filled with man-made interference down here. We are looking for satellite techniques to take us away from the mess we have made down here, and I hope that a similar confusion will not arise in the case of the 21-cm band. The wave band that we should preserve unspoilt goes from 21 cm down to 35 or 40 cm, the wavelength at which 21-cm radiation will be seen, Doppler-shifted, from the distant parts of the expanding universe when the new sensitive receivers are used. The preparations must be made now, before the problem arises. If a lot of equipment were constructed that crowded out this band, it would be extremely hard at a later date to get the range clear again. I understand, of course, that wave bands are of great value to technology and commerce, but so is the progress of fundamental research.

Index

Aarons, J., 81
Alfvén waves, 126, 128
Allen, E. W., 99
Appleton, Sir Edward, 12
Aristotle, 7
Aurora: early descriptions, 7; geometry of radar reflections, 15–23; micropulsations, 118; noise generation, 38; radio and radar observations, 11–13, 23–31; reflection of radio waves, 8f; visual and optical studies, 9–11; zones of maximal frequency, 8
Auroral echoes, 13–14; aspect sensitivity, 32, 35, 38, 39; diffuse, 26, 29, 30; Doppler shifts, 28, 31, 33, 34; frequency dependence, 32, 35–38, 39, 40; intensity, 36, 37; ionospheric disturbances, 27; magnetic disturbances, 27; scattering mechanism, 32, 34–38; solar disturbances, 40; sporadic-E, 39
Auroral spectrograph, 10
Auroral zone, 8, 9, 12; optical, 21–23; radar, 21–23
Autler, S. H., 46
Automobile ignition noise, 3

Beamwidth corrections, 141, 156–158; scale factors, 148; sharpening, 149; smoothing, 143–148
Booker, H. G., 34, 35, 36; auroral scattering mechanism, 34–38
Bracewell, R. N., 139

Cassiopeia A, 45, 88, 159, 173
Centaurus A, 159, 173
Cerenkov radiation, 39
Charged particles from sun: whistlers, 98, 99, 123f
Communication interference: aurora, 8f; cosmic noise, 151f, 172f; man-made noise, 1–6

Cosmic radio noise sources, 151f; horizon coordinate grids, 164–168; maps, 151, 158, 160–161, 163; nomogram, 169–170, 171; published surveys, 154–155; spectral law, 163
Cosmic rays: interplanetary magnetic fields, 123, 126–127; solar, 108, 123
Cygnus A, 43, 44, 45, 159, 173; scintillation, 44, 45

Discrete-Interval theorem, 142
Discrete noise sources, 159, 172–175; Cassiopeia A, 45; Cygnus A, 43, 44, 45; scintillation, 43–46
Doppler shift: auroral echoes, 28, 31, 40; meteor trails, 65; solar whistlers, 125
D-region, 99

Earth: magnetic field, 111–112; pulsations of field, 124, 127
E-layer, 31, 36, 115; auroral echoes, 26; auroral reflections, 17; whistlers, 94–95
Eshleman, V. R., 92

FCC, 3, 4, 5
F-layer, 115
Fritz, H., 9

Gartlein, C. W., 10
Geminid meteor stream, 79, 92
Geomagnetic disturbance: whistlers, 98, 99 (See also Magnetic storms)
Geomagnetic poles, 8
Goldberg, P. A., 92, 99, 111, 120
Green, P. E., 138
Gruber, S., 46, 47
Gyro-frequency of ions, 119, 128

Harvard Radio Astronomy Station, Fort Davis, 101; equipment, 101–105